MIXED-MODE SIMULATION
AND
ANALOG MULTILEVEL SIMULATION

THE KLUWER INTERNATIONAL SERIES IN ENGINEERING AND COMPUTER SCIENCE

VLSI, COMPUTER ARCHITECTURE AND DIGITAL SIGNAL PROCESSING

Consulting Editor

Jonathan Allen

Other books in the series:

PIPELINED ADAPTIVE DIGITAL FILTERS, Naresh R. Shanbhag, Keshab K. Parhi
ISBN: 0-7923-9463-1

TIMED BOOLEAN FUNCTIONS: A Unified Formalism for Exact Timing Analysis, William K.C. Lam, Robert K. Brayton
ISBN: 0-7923-9454-2

AN ANALOG VLSI SYSTEM FOR STEREOSCIPIC VISION, Misha Mahowald
ISBN: 0-7923-944-5

ANALOG DEVICE-LEVEL LAYOUT AUTOMATION, John M. Cohn, David J. Garrod, Rob A. Rutenbar, L. Richard Carley
ISBN: 0-7923-9431-3

VLSI DESIGN METHODOLOGIES FOR DIGITAL SIGNAL PROCESSING ARCHITECTURES, Magdy A. Bayoumi
ISBN: 0-7923-9428-3

CIRCUIT SYNTHESIS WITH VHDL, Roland Airiau, Jean-Michel Berge, Vincent Olive
ISBN: 0-7923-9429-1

ASYMPTOTIC WAVEFORM EVALUATION, Eli Chiprout, Michel S. Nakhla
ISBN: 0-7923-9413-5

WAVE PIPELINING: THEORY AND CMOS IMPLEMENTATION, C. Thomas Gray, Wentai Liu, Ralph K. Cavin, III
ISBN: 0-7923-9398-8

CONNECTIONIST SPEECH RECOGNITION: A Hybrid Appoach, H. Bourlard, N. Morgan
ISBN: 0-7923-9396-1

BiCMOS TECHNOLOGY AND APPLICATIONS, SECOND EDITION, A.R. Alvarez
ISBN: 0-7923-9384-8

TECHNOLOGY CAD-COMPUTER SIMULATION OF IC PROCESSES AND DEVICES, R. Dutton, Z. Yu
ISBN: 0-7923-9379

VHDL '92, THE NEW FEATURES OF THE VHDL HARDWARE DESCRIPTION LANGUAGE, J. Bergé, A. Fonkoua, S. Maginot, J. Rouillard
ISBN: 0-7923-9356-2

APPLICATION DRIVEN SYNTHESIS, F. Catthoor, L. Svenson
ISBN:0-7923-9355-4

ALGORITHMS FOR SYNTHESIS AND TESTING OF ASYNCHRONOUS CIRCUITS, L. Lavagno, A. Sangiovanni-Vincentelli
ISBN: 0-7923-9364-3

HOT-CARRIER RELIABILITY OF MOS VLSI CIRCUITS, Y. Leblebici, S. Kang
ISBN: 0-7923-9352-X

MOTION ANALYSIS AND IMAGE SEQUENCE PROCESSING, M. I. Sezan, R. Lagendijk
ISBN: 0-7923-9329-5

HIGH-LEVEL SYNTHESIS FOR REAL-TIME DIGITAL SIGNAL PROCESSING: The Cathedral-II Silicon Compiler, J. Vanhoof, K. van Rompaey, I. Bolsens, G. Gossens, H. DeMan
ISBN: 0-7923-9313-9

MIXED-MODE SIMULATION AND ANALOG MULTILEVEL SIMULATION

Resve Saleh

Shyh-Jye Jou

University of Illinois

A. Richard Newton

University of California

SPRINGER SCIENCE+BUSINESS MEDIA, LLC

Library of Congress Cataloging-in-Publication Data

Saleh, Resve A., 1957-
Mixed-mode simulation and analog multilevel simulation / Resve Saleh, Shyh-Jye Jou, A. Richard Newton.
p. cm. -- (Kluwer international series in engineering and computer science)
Includes bibliographical references and index.
ISBN 978-1-4419-5144-1 ISBN 978-1-4757-5854-2 (eBook)
DOI 10.1007/978-1-4757-5854-2
1. Computer simulation. I. Jou, Shyh-Jye. II. Newton, A. Richard (Arthur Richard), 1951- . III. Title. IV. Series.
QA76.9.C65S26 1944
621.3815'01'13--dc20 94-20345
CIP

Printed on acid-free paper.

TABLE OF CONTENTS

PREFACE

Our purpose in writing this book was two-fold. First, we wanted to compile a chronology of the research in the field of mixed-mode simulation and analog multilevel simulation over the last ten to fifteen years. A substantial amount of work was performed during this period of time but most of it was published in archival form in Masters theses and Ph.D. dissertations. Since the interest in mixed-mode simulation and analog multilevel simulation is growing, and a thorough review of the state-of-the-art in the area was not readily available, we decided to publish the information in the form of a book.

Secondly, we wanted to provide enough information to the reader so that a prototype mixed-mode simulator could be developed using the algorithms in this book. The SPLICE family of mixed-mode programs is based on the algorithms and techniques described in this book and so it can also serve as documentation for these programs.

In this new edition of the book, we have added a substantial amount of information on the mixed-mode interface in Chapter 7 and automatic mixed-mode partitioning in Chapter 8. We have also improved the review of existing mixed-mode simulators so that the reader is better able to select the most appropriate one for their application. Chapter 9 is a new chapter on analog multilevel simulation. The iMACSIM program, developed at the University of Illinois, is based on the contents of this chapter, so it serves as documentation for this program.

Although, there are some omissions of other relevant research work in this book, space limitations did not allow us to include everything. However, some of the other research has already been published by Kluwer Academic Publishers and others, and we wanted to avoid any duplication.

ACKNOWLEDGEMENTS

The authors dedicate this book to Prof. D. O. Pederson for inspiring this research work and for providing many years of support and encouragement. The authors enjoyed many fruitful discussions and collaborations with Jim Kleckner, Young Kim, Alberto Sangiovanni-Vincentelli, Jacob White and Jaidip Singh, and we thank them for their contributions. Jaidip Singh, Tom Thatcher, Victor Ma, Dave Overhauser and Narendra Jain also provided useful contributions to the new version. We also thank the countless others who participated in the research work and read early versions of this book. Lillian Beck proofread the text and provided many suggestions to improve the manuscript. Yun-Cheng Ju did the artwork for the most of the illustrations. Mei-Hsin Wu assisted with the text and new illustrations for this edition of the book. The second version was also reviewed by Uma Ekambaram, Luis Amaya and Brian Antao. Jaidip Singh and Xiaocun Xu prepared most of the new simulation results provided in the book.

The Semiconductor Research Corporation provided a substantial amount of funding for the body of work presented in this book. We gratefully acknowledge their continuing support for research and education in this area. Other funding and computer equipment was provided by the Natural Sciences and Engineering Research Council (NSERC) of Canada, the Hewlett-Packard Company, Toshiba Corporation, Digital Equipment Corporation, Analog Devices and Motorola Inc., and we appreciate their contributions.

CHAPTER 1

INTRODUCTION

1.1. THE SIMULATION PROBLEM

Computer simulation is used in a variety of different fields to predict the behavior of physical systems whenever it is inappropriate, or too expensive, to build the actual system to observe its behavior. In electrical engineering, circuit simulation is used routinely in the design of integrated circuits (IC) to verify circuit correctness and to obtain detailed timing information before an expensive and time-consuming fabrication process is performed. In fact, it is one of the most heavily used computer-aided design (CAD) tools in terms of CPU-time in the IC design cycle. The popularity of this form of simulation is primarily due to its reliability and its ability to provide precise electrical waveform information for circuits containing complex devices and all associated parasitics.

Detailed circuit simulation has been used extensively for IC design since the early 1970s. However, the ever-increasing number of devices on a single silicon chip has led to development of a number of higher-level simulation tools to cope with the complexity of the problem. In digital IC design, these tools include behavioral simulators, register-transfer level (RTL) simulators, gate-level logic simulators, and switch-level simulators. These programs have been used to verify circuit functionality and to obtain first-order timing characteristics. Typically, the higher-level tools provide enough information to design working circuits. However, there is still a significant time lag between a functioning circuit and a circuit which meets the design specifications - particularly in the case of high-performance custom integrated circuits. In fact, circuit simulation is the only tool which provides enough detail to ensure that circuits of this type will meet specifications over a wide range of circuit parameters and operating conditions.

At the present time, the most popular circuit simulation tool is the SPICE2 program [NAG75]. There are many thousands of copies of this program in use, as well as a number of versions of "alphabet-SPICE" (e.g., HSPICE, PSPICE, IGSPICE) being marketed commercially. All of these programs offer a wide variety of analyses including dc analysis, time-domain transient analysis, ac analysis, noise analysis and distortion analysis. Of these, the time-domain transient analysis is the most computationally expensive in terms of CPU-time. The SPICE program was originally designed to simulate circuits containing up to 100 transistors. However, at certain companies, this program is often used to simulate circuits containing over 10,000 transistors at great expense! The program is accessed over 50,000 times per month at some of companies with a "job mix" that conforms to the 80-20 rule. That is, 80% of the SPICE runs are on small circuits which consume only 20% of the total CPU-time used each month, while 20% of the jobs are very large and consume 80% of the CPU-time used each month. Today, designers require simulators that can accommodate digital circuits with 500,000 to 1,000,000 transistors in a single run. Therefore, the development of fast but accurate simulation methods for very large-scale integrated (VLSI) circuits continues to be an important area of research and development.

With the advent of submicron technologies and faster clock speeds, complex VLSI systems including both analog and digital circuits have been integrated onto a single chip. It is estimated that roughly 60% of all semi-custom ICs includes some analog circuitry, and 30% of all standard cell designs include analog functions that account for 10% of the chip area. The percentage is even higher for custom ICs such as voltage regulators, phase-locked loops, filter codecs, line equalizers, ISDN line interface circuits, line cards, DSP with on board A/D and D/A, speech recognition circuits, FDM transducers, power up/down sense circuits, and dial tone/pulse generators. This proliferation of *mixed-signal ICs* has lead to problems in almost every area of IC development including design, layout, fabrication, testing, and manufacturing.

The trend of integrating complete analog/digital systems on a single chip has also resulted in new challenges in modeling and simulation. First of all, problems in analyzing these circuits arise due to the different modeling and simulation approaches used for analog and digital circuits, and differences in their accuracy requirements. The different approaches must be combined in a consistent manner so that entire systems can be simulated with the appropriate speed/accuracy tradeoff. Secondly, system-level design is being emphasized to cope with the complexities of large designs. In this context, behavioral modeling and simulation are essential to the validation of a proposed architecture before a detailed design begins. This has been used in digital system design for many years with great success. Designs are described and simulated at the behavioral level using standard hardware description languages. Unfortunately, in thc analog domain, circuits are still designed and verified at the electrical level, despite the fact that circuit simulation is very time-consuming. For mixed-signal simulation, it is essential to provide an analog modeling and simulation environment similar to the digital domain, that would allow the designer to model components at the behavioral level and then perform system-level analog simulation. With this capability, the overall architecture of a proposed mixed-signal design could be verified at a high-level in a reasonable amount of time. After the verification process, the detailed design could be performed based on specifications derived from the high-level simulation.

This book addresses the problems of simulating entire mixed analog/digital systems in the time-domain. A complete hierarchy of modeling and simulation methods for analog and digital circuits is described. Fig. 1.1 lists all the levels in the hierarchy for the two domains. These levels are briefly described in the sections to follow as a motivation for techniques developed in the rest of the book.

1.2. LEVELS OF SIMULATION FOR DIGITAL CIRCUITS

This section provides an overview of the different levels of simulation that have been used in digital circuit design. It begins by describing the most detailed level and then moves to progressively higher and higher levels of abstraction. Each level is illustrated in Fig. 1.2 for a CPU design.

Digital	*Analog*
Behavioral	*Behavioral*
RTL / Gate	*Ideal Functional*
Switch Level	*Non-Ideal Functional*
Electrical	*Electrical*

Figure 1.1: Levels of Simulation

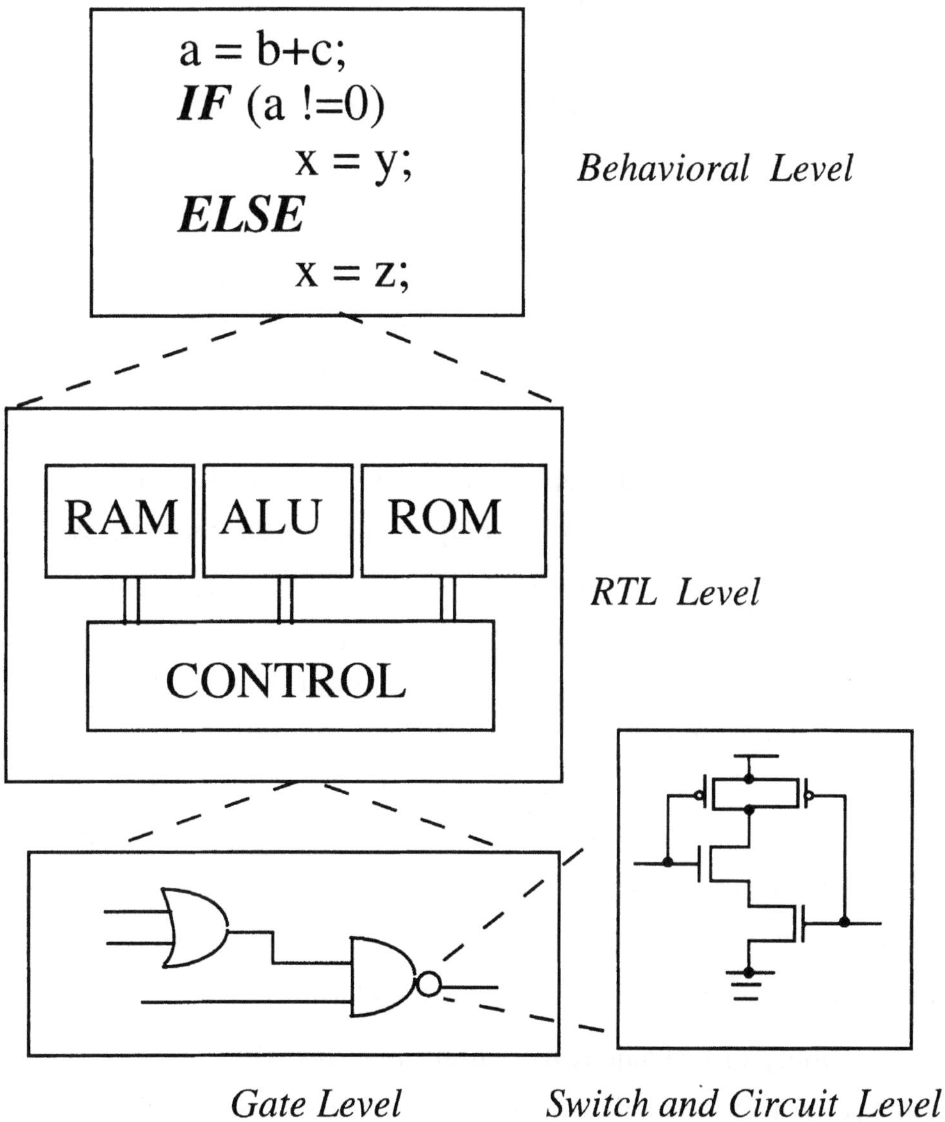

Figure 1.2: Levels of Abstraction in Digital Simulation

1.2.1. Electrical Simulation

Electrical or circuit level simulation provides the greatest amount of detail. The electrical transient analysis problem in SPICE involves the solution of a system of nonlinear, first-order, ordinary differential equations. These equations model the dynamic characteristics of the circuit for a set of applied input voltages and initial conditions. The solutions are voltage waveforms *across* pairs of circuit nodes and current waveforms *through* circuit elements. Usually the designer is interested in only a subset of the entire set of solutions.

Standard circuit simulators use *direct methods* to solve the circuit equations. Briefly, direct methods employ some form of numerical integration to convert nonlinear differential equations into a set of nonlinear difference equations. These equations are solved simultaneously using the iterative Newton-Raphson method. This involves a conversion of the nonlinear equations into linear equations, and their subsequent solution using a sparse LU decomposition technique [NAG75]. There are two limitations in this approach that make it inappropriate for large circuits. One fundamental problem is that the sparse linear solution dominates the run time for large circuits [NEW83]. The second limitation is that, at each time point, *all* the variables in the system are solved using a common time-step based on the fastest changing component in the system. This can be inefficient for both small and large circuits, but it is more significant for very large problems where most of the components are either changing very slowly or not changing at all.

A variety of techniques have been investigated to improve the performance of circuit simulators. Early work in this area includes *timing simulation* [CHA75, NEW78, DEM80], which is a simplified form of relaxation-based circuit simulation, and *tearing methods*, which have been applied at both the linear [SAN77, YAN80, SAK81] and nonlinear

[RAB79] equation levels. More recently, the relaxation-based approaches have been the focus of intensive research. In particular, the Waveform Relaxation method [LEL82, WHI83] has been implemented in a number of programs including RELAX [LEL82, WHI83], SWAN [DUM86], TOGGLE [HSI85], RealAx [MAR85], MOSART [CAR84] and iDSIM [OVE89]; and Iterated Timing Analysis [KLE83, SAL84], based on non-linear relaxation, has been implemented in SPLICE [SAL83, KLE84, ACU89], ELDO [HEN85] and SISYPHUS [GRO87].

1.2.2. Gate-Level Simulation

When the complexity of an integrated circuit design reaches the point at which electrical analysis is no longer cost effective, logic or *gate-level* simulation can be used [BRE76]. In logic simulation, transistors are usually grouped into logic *gates* and modeled at the gate level. This form of simplification, sometimes referred to as *macromodeling*, can result in greatly enhanced execution speed by reducing the number of models to be processed and simplifying the arithmetic operations required to process each transistor group. Rather than dealing with voltages and currents at signal nodes, discrete logic *states* are defined, and simple Boolean operations are used to determine the new logic value at each node. Boolean operations are generally the most efficient operations available on a digital computer.

A logic simulator that uses event-driven, selective-trace techniques is typically 100 to 1000 times faster than the *most efficient* forms of electrical analysis. It can also provide first-order timing information, including the detection of hazards, glitches, and race conditions. In addition, it can provide output information regarding any illegal states or conflict conditions that may arise at any node in the circuit. The number of logic states used in a simulation, their meaning, the logic delay models and the scheduling algorithm all have a profound impact on both the speed and accuracy of the results. It is this wide variety of factors that has resulted in the development of such a large number of logic simulators - almost every one addressing a different set of tradeoffs.

1.2.3. Switch-Level Simulation

Recently, switch-level simulation [BRY80, RAO89] has become the preferred form of logic simulation for MOS digital circuits. In this approach, the circuit is entirely simulated at the transistor level, rather than at the gate level. The transistors are modeled as gate-controlled switches and operate as follows: if the transistor is "ON," it is viewed as a closed switch and it may transfer a signal value from one node to another; if the transistor is "OFF," it is viewed as an open switch and is incapable of transmitting any signals through it. The network is composed of a set of nodes connected by these switches, and the logic value at each node is determined using this idealized transistor model. Usually a strength is associated with each transistor switch when in the closed position to model the conductance of the device. This strength is used to determine the effective conductance of signal paths from any node to the supply and ground nodes. The capacitance at each node can also be modeled using a node strength that is proportional to the size of the capacitance. Many of the important features of MOS circuits, such as charge-sharing and bidirectionality, can be modeled using this switch-level model, although detailed timing and voltage level information are not usually provided.

A number of researchers have attempted to incorporate timing information at the switch level at the cost of additional CPU-time. Simulators that fall into this category are MOTIS [CHE84], RSIM [TER83], ELOGIC [KIM84], SPECS [DEG84], MOSTIM [RAO85], CINNAMON [VID86], SPECS2 [VIS86] and iDSIM [OVE88]. Programs such as RSIM treat

MOS transistors as linear resistors and compute signal transition times using the Penfield-Rubenstein technique [PEN81], which is an RC-delay modeling technique. Although the method is extremely efficient, the overall accuracy of this approach is limited due to the simplified nature of the delay modeling. MOSTIM and iDSIM determine the delay directly using lookup tables that are generated during a pre-characterization phase for recognizable transistor configurations. These tables account for factors such as device sizes, loading and input slew-rate. The ELOGIC and SPECS programs compute the delays by using electrically-based table lookup device models. The waveforms are generated as piecewise linear segments using the computed delays. Both approaches provide for variable precision by allowing the user to specify the number of voltage or current levels to be used in the simulation.

1.2.4. Register-Transfer Level Simulation

Register-Transfer Level (RTL) [BRE75] simulation is concerned with logic circuits described at a higher level of abstraction. Combinational components, (such as gates, multiplexers, decoders, encoders, adders, and arithmetic units) and sequential components (such as registers and counters) may be used in RTL simulators. RTL simulation has been used extensively for data path design. It is used for both the description and simulation of the designs when evaluating alternative architectures. The set of statements describing the circuit operation involves a sequence of register transfers and arithmetic operations that are similar to data-flow descriptions. In the description, related bits of information are usually grouped into ordered sets of words or buses for convenience and to establish logical relationships. Although RTL simulators are widely used to design computers, they do not usually provide information regarding races, hazards, illegal states or critical timing constraints.

1.2.5. Behavioral Level Simulation

Behavioral level simulators [HIL80, INF84, INS84] allow the designer to define arbitrary functional blocks, both combinational and sequential, that can be used in system-level simulation. Two types of blocks may be defined: *structural* and *behavioral*. Structural blocks describe how a number of functional blocks are interconnected. A behavioral block contains a detailed description of the operations to be performed on the inputs to produce the outputs of the block. The statements describing the operations are usually written in a high-level language, typically a hardware description language (HDL), and then translated to a standard programming language format and compiled into the simulator. Within the digital arena, *VHDL* (VHSIC HDL) [IEE88] and *Verilog* [THO91] are fairly mature, well-accepted industrial standards. When a behavioral simulator is executed, the operations of the system are emulated. Examples of applications that are appropriate for simulation at the behavioral level are: verifying the system timing of a CPU; checking a proposed network protocol for a local-area network; and validating the operations in DMA controller sequence.

1.3. LEVELS OF SIMULATION FOR ANALOG CIRCUITS

The various levels of simulation discussed thus far are shown on the left side of Fig. 1.1. In defining this hierarchy of methods, there was a clear bias towards digital circuits, since they usually dominate most of the mixed-signal chip area, and tend to profit greatly from hierarchical representation and the use of the higher levels of simulation. However, as shown on the right side of the figure, there is also a corresponding set of levels in the analog domain that has been overlooked until recently. The various levels in the analog hierarchy are described in the following sections using the sampled-data filter of Fig. 1.3.

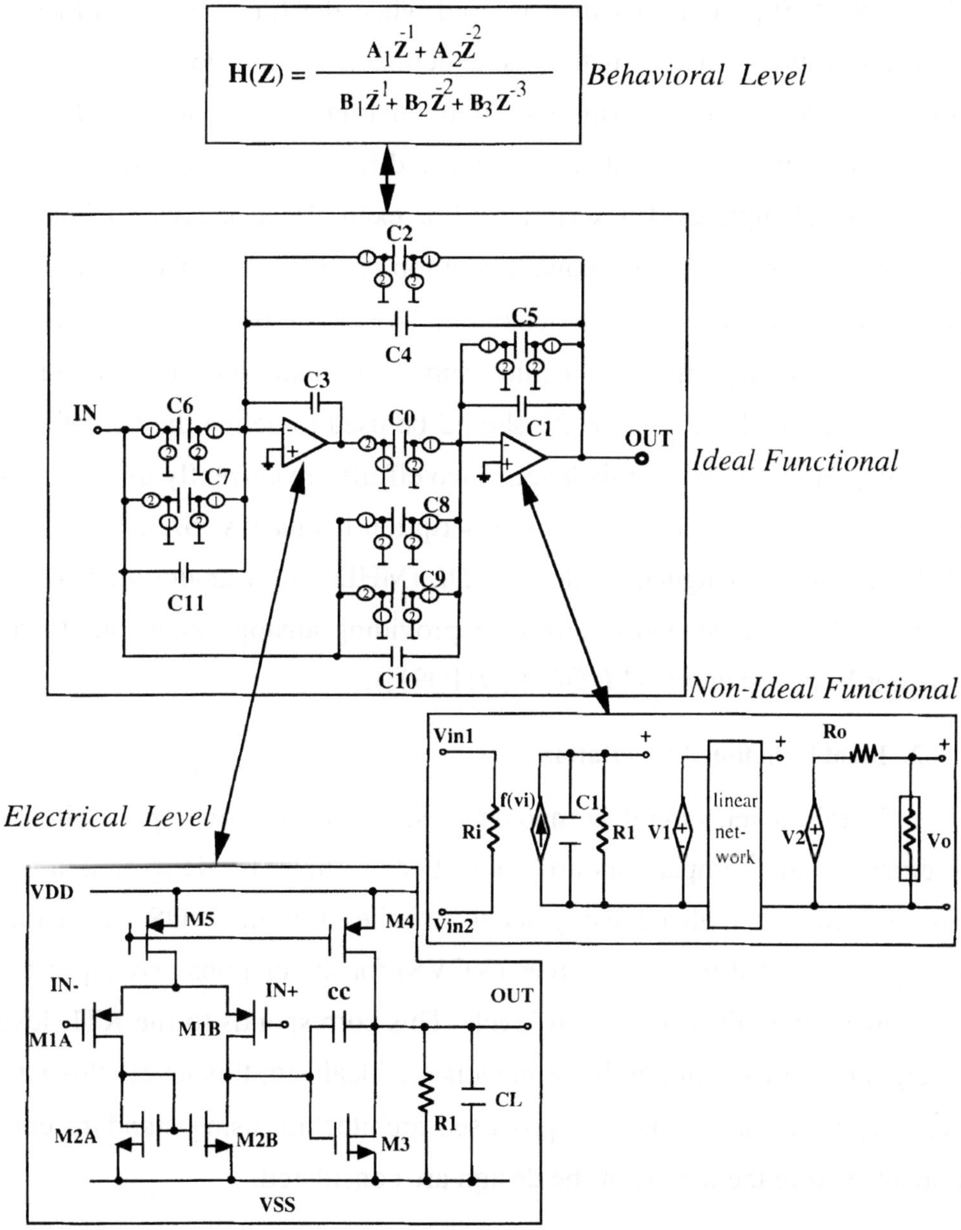

Figure 1.3: Levels of Abstraction in Analog Simulation

1.3.1. Behavioral Simulation

At the highest level is behavioral simulation [CHA92, GIE92, SIN91, RUM89, VLA90] which is used when the function of a block is known but its detailed structure is undefined. At this level, individual blocks are described in terms of **s**-domain transfer functions, **z**-domain transfer functions, differential equations, difference equations, or some other form of high-level description. For example, a behavioral level **z**-domain transfer function is shown at the top of Fig. 1.3 for a discrete-time filter. The interaction between behavioral blocks can be described in terms of signal-flow diagrams that include summers, multiplers, differentiators and integrators. Ideally, an HDL should be used to define the behavior of an analog circuit block at this level. Two efforts are currently underway to develop standard analog hardware description languages [SAL94A]. One of these is a new language called *MHDL* [MHD91] for analog/microwave circuits while the second is aimed at providing analog extensions to the existing VHDL standard (*VHDL-A* [VHD91]).

1.3.2. Ideal Functional Simulation

At this level, circuit components, such as ideal opamps, switches, integrators, and comparators are used. For example, the **H**(**z**) function in Fig. 1.3 can be realized using ideal switches for the MOS transistors, voltage-controlled voltage sources (VCVS) for the opamps and capacitors as shown at the ideal functional level. This corresponds to the RTL level in digital circuits. Although the models are idealized, this level allows the designer to quickly validate a proposed architecture using standard components before the details of the design are considered.

1.3.3. Non-Ideal Functional Simulation

This level is similar to the ideal functional level except that the first-order and second-order details are included in the models. Non-ideal function blocks are commonly referred to in the literature as *macromodels*. They are usually constructed by connecting a number of primitive elements together, each representing some inherent feature of the block being macromodeled. Relationships between the components within the macromodel may be defined using linear and nonlinear dependent sources such as voltage-controlled voltage sources (*VCVS*), current-controlled voltage sources (*CCVS*), current-controlled current sources (*CCCS*), and voltage-controlled current sources (*VCCS*). Complex models that include nonlinear properties, dynamic behavior, limiting effects and detailed input/output characteristics may be provided at this level. For example, in Fig. 1.3, piecewise-linear gain, finite bandwidth and input/output resistance have been included in the macromodel of the operational amplifiers. The capacitances and resistances could also be included in the MOS switches when simulating the switched-capacitor filter circuit. This level roughly corresponds to the gate and switch levels in the digital domain that includes timing information and other electrical effects.

1.3.4. Electrical Simulation

Finally, at the most detailed level, electrical simulation is available, which corresponds directly to the same level on the digital side of the Fig. 1.1. Here, the operational amplifiers would be represented in terms of their MOS transistors with detailed models as shown in Fig. 1.3. In addition to time-domain simulation, some frequency-domain analysis capabilities such as ac analysis, sensitivity, noise and distortion must be available to the analog designers. However, these topics will not be addressed in this book. The interested reader is encouraged to consult [KUN90] for more information on this subject.

1.4. MIXED-MODE AND ANALOG MULTILEVEL SIMULATION

The levels of simulation described in the previous two sections are listed in Tables 1.1 and 1.2 from the highest level of abstraction to the lowest level. The relative runtime cost and accuracy of each digital simulation level is provided in Table 1.1 for the hypothetical simulation of a 32-bit microprocessor. The corresponding example for the analog levels is a fifth-order switched-capacitor filter. Although this is a small example, it is useful for our purpose as it produces extremely long runtimes, and can easily be represented at each level in the hierarchy. The reader should notice that the progression from the behavioral level to the electrical level provides an increase in the accuracy of the simulation at the cost of more CPU-time. A progression in the opposite direction allows larger and larger circuits to be simulated for a given amount of CPU-time, or requires less and less CPU-time to simulate a given circuit. However, each level uses less precision in its signal representation. This often translates to less

Level	Relative Cost	Capability and Accuracy
Behavioral (B)	1	Algorithmic verification, some timing information
RTL (R)	10	Functional verification, some timing information
Gate (G)	100	Functional verification, first-order timing information
Switch (S)	1000	Functional verification, first-order timing information
Timing (T)	10000	Detailed waveform information with variable accuracy
Electrical (E)	1000000	Most accurate form of simulation

Table 1.1: Relative Cost and Accuracy of Digital Simulation

Level	Relative Cost	Capability and Accuracy
Behavioral (B)	1	System level verification
Ideal Functional (I)	10	Functional verification, 1st-order electrical information
Non-Ideal Functional (N)	100	Functional verification, 1st-order and 2nd-order electrical information
Electrical (E)	1000	Most accurate form of simulation

Table 1.2: Relative Cost and Accuracy of Analog Simulation

accuracy in the results due to modeling limitations.

There are many situations for which one level of simulation is not sufficient for the simulation of an entire design. One common situation arises in the design of a mixed analog and digital circuit. Logic simulators do not generally have the capability to model analog circuitry, and it is usually too expensive to simulate the entire mixed-signal design in a circuit simulator. In this case, it would be convenient if a simulator that included both electrical and logic simulation capabilities were available.

Multiple levels of abstraction are commonly used in "top-down" or "bottom-up" design styles. In both cases, the entire design at any given point in time may be represented at a number of different levels of abstraction. One designer may have written the behavioral specification of his/her portion of the design, while a second is completing the detailed gate-level design, and a third is performing transistor-level cell library development. Furthermore, a designer often uses multiple levels of abstraction in a schematic diagram to convey the important aspects of the design as shown in Fig. 1.4 for a portion of a digital control circuit. To ensure that a design represented schematically at many levels in the hierarchy is functionally correct at any stage of the design process, a simulator that handles all possible levels of abstraction would be extremely useful.

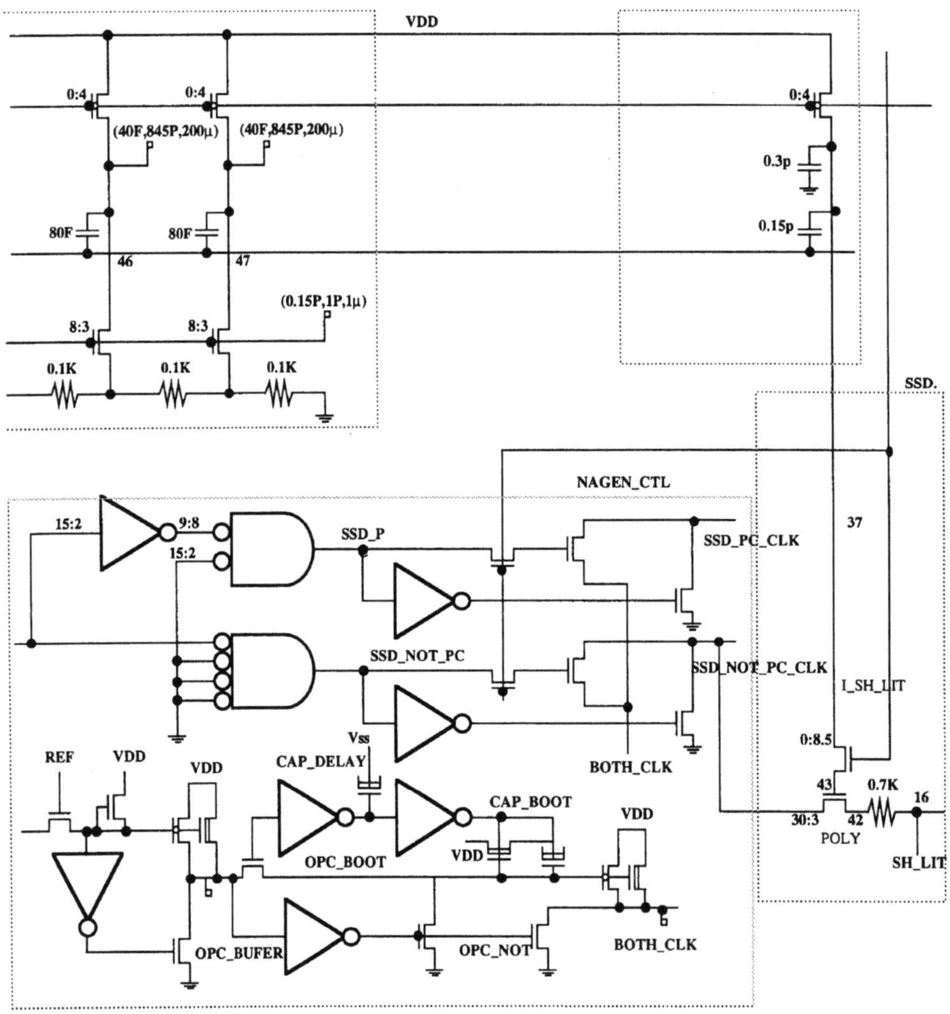

Figure 1.4: A Portion of a Digital Control Circuit

Multilevel simulation can also be used for the purpose of accurate circuit modeling. For example, standard gate-level simulators are not capable of simulating the behavior of certain properties of MOS digital circuits such as bidirectionality and charge-sharing. Therefore, the mixing of switch-level simulation and gate-level simulation would provide an effective balance between simulation speed and functional accuracy. On the other hand, the idealized transistor model in switch-level simulation is not appropriate for the simulation of certain pass transistor configurations, and other circuits where the device W/L ratios are important. For these cases, mixing electrical-level, switch-level and gate-level simulations would be useful.

All of the situations cited above require a simulator that allows different portions of the circuit to be described and simulated at different levels of abstraction. That is, where accuracy is not a critical issue, higher levels of simulation can be used, but where proper modeling of the circuit is a problem or detailed timing information is desired, the lower levels of simulation can be used. CAD tools that address this need are referred to as *multilevel simulators* or mixed-level simulators. Clearly, the most important issues in multilevel simulators arise when combining the gate-level simulation with electrical-level simulation. This is referred to as *mixed-mode simulation* and this is the main topic of this book. When the term mixed-mode simulator is used in general, it usually refers to a simulator that can simulate the digital part of a mixed-signal design using any of the levels in the digital hierarchy shown in the left side of Fig. 1.1 while analog part of the design uses the electrical level.

A recent effort in analog simulation is to develop a multilevel simulation environment that incorporates the different levels of simulation listed in the right side of Fig. 1.1. It allows the designer to represent different portions of the design at any desired level in either the time-domain or the frequency-domain and then choose the desired form of analysis. For example, the designer could model a discrete-time linear filter using a behavioral **z**-domain transfer function representation, and simulate a multilevel description of the circuit, with some components represented at the transistor level and others at the behavioral level. The **z**-domain model can be processed much more cheaply than the equivalent amount of electrical-level circuitry. Once the performance of the overall system has been verified, the detailed design of the filter can be performed at the transistor level.

Great progress has been made in the development of analog multilevel simulators but much work still remains, especially in the area of behavioral simulation and the development of nonideal functional models to include nonlinearities and noise effects. In addition, the issues of maintaining consistency when models from different domains are connected to the same node and performing the necessary transformations are still under investigation. As a final step in the evolution, the analog simulation levels should be combined with the digital simulation levels so that both of the hierarchies in Fig. 1.1 can be mixed and matched easily within a single simulation environment. The combination of these techniques would produce a powerful tool that could address most of the problems of mixed-signal simulation.

This book presents a unified approach to simulation that allows both mixed-mode simulation and analog multilevel simulation to be combined in the same environment. The primary focus is mixed-mode simulation in which circuit simulation is combined with logic simulation. This book also addresses analog multilevel simulation which is based on a similar hierarchy of modeling and simulation methods for analog circuits. The research work in mixed-mode simulation and analog multilevel simulation over the last 10-15 years is described in detail. This book identifies the key advances that have been made in these two areas including: event-driven electrical simulation, signal mapping across the mixed-mode interface, automatic mixed-mode partitioning, analog behavioral modeling and simulation and mixed-domain simulation. The algorithms provided in this book allow the reader to implement prototype analog and mixed-signal simulators.

1.5. BASIC ISSUES IN MIXED-MODE SIMULATION

This section describes the issues involved in designing a mixed-mode simulator. The specific issues of analog multilevel simulation are postponed until Chapter 9, although many of the points raised in this section also apply to analog multilevel simulation. The mixed-mode issues are as follows:

Choice of Simulation Levels: First, and foremost, is the issue of what types of simulation to include in the simulator. This depends on the intended application of the simulator. If the design is primarily digital in nature, the combination of gate, RTL and behavioral simulations would be appropriate. For MOS designs, it may be better to incorporate gate and switch-level simulations. For mixed-signal circuit designs containing both MOS and bipolar transistors, it may be necessary to mix gate, switch and electrical level simulations. Ideally, one would prefer to combine all the levels of simulation into one program, but the development time would be significant.

Simulator Architecture: A mixed-mode simulator must be flexible and extensible so that algorithms and device models can be added or removed easily as the technology and the simulator requirements evolve. An appropriate choice of simulator architecture is necessary to achieve this goal. The architecture described in this book is based on the use of the event-driven, selective-trace paradigm at all levels of abstraction. This permits the exclusive simulation of activity, and it is a necessary feature when simulating large digital systems. It is also consistent with the algorithms commonly used in most simulators, except for standard electrical simulators, which must be modified to fit within the event-driven framework.

Event Definition and Event Scheduling: To establish event-driven, selective-trace simulation, the notion of an *event* must be defined at each level in the simulator. An event is a change in state of some node in the circuit that may affect other components in the circuit. The effect of an event is to cause all fanout components to be processed, and possibly new events to be scheduled, if changes in their output nodes occur. A key issue in mixed-mode simulation is to define an event scheduling policy between different modes of simulation.

Mixed-Mode Interface: A consistent representation for signals over all simulation levels is critical for accurate mixed-mode simulation. In the higher levels of simulation, the signal value is usually represented using hexadecimal values for collections of bits or logic values. At the other extreme, electrical simulation uses 64-bit double precision words to represent real values of voltage. A mixed-mode simulator must be able to manage these different signal types at the interface and map them from one representation to another without a significant loss in accuracy.

Representation of Time: Time is usually represented as a real number in electrical simulation and as an integer in logic and higher level simulations. Typically the time steps chosen in electrical simulation are very small (order of nanoseconds to picoseconds depending on the time constants associated with the devices and parasitic elements), whereas in the logic level and higher levels of simulation, it is usually an integer multiple of some basic unit of time that depends on the clock period or the delays of the logic gates. This disparity between the various representations of time must also be resolved in the mixed-mode environment.

Partitioning: Circuit partitioning is a key factor in obtaining efficiency and accuracy from mixed-mode simulation. The main question is to determine which portions of the circuit must be simulated at the most detailed level and which portions will profit from simulation at higher levels of abstraction with an acceptable loss in accuracy. The prospects of performing this task automatically seem formidable, especially in the case of large circuits. To date, most of the simulators available require that circuit designers be responsible for the partitioning process, since they are familiar with the nature of the design.

User Interface: Another important consideration when designing a mixed-mode simulator is the user interface to the simulator. The interface must be graphics-oriented, highly interactive, and provide the features of schematic capture, simulation control and output post-processing. While a variety of schematic packages with these features do exist commercially, there are a number of additional requirements in mixed-mode simulation. First, the front-end must allow a hierarchical representation of the circuit in which each successive level of the hierarchy implies a different level of abstraction. That is, each level in the hierarchy represents a different form of simulation in the associated mixed-mode simulator. This implies that all of the components representing the circuit at two or more different levels must have the same functional behavior to guarantee correct results. Therefore, some convenient way of verifying the consistency of different representations of the same circuitry must be provided. The capability of adding new components, specifically macromodels or high-level RTL and behavioral models, should be made simple.

Mixed-mode Timing Control: Time-step synchronization between the analog and digital simulators due to local and global feedback loops is an important factor in determining the accuracy and speed of mixed-mode simulation. Some methods to synchronize the time of simulation at the analog/digital interface are lock-step, digital-controlled, analog-controlled and roll-back schemes.

In the lock-step scheme, the step sizes are determined by simulator requiring the smaller time-step. Typically, the analog simulator sets the step size and the digital simulator is forced to use these values. The advantage of this method is that synchronization will always be maintained since digital and analog portions are simulated at the same time, but the disadvantage is that latency in digital simulation can not be exploited fully and this degrades the overall simulation speed.

In the digital-controlled scheme, the digital signals are used as inputs to the analog simulators. Therefore, the analog simulator is forced to "catch-up" to the next digital event. Accuracy of the overall simulation is usually compromised using this approach. On the other hand, some simulators perform all levels of simulation in the continuous-time domain. In this case, all signals are represented as analog voltages and, therefore, the analog simulator in charge of the time-step control. This is called analog-controlled time synchronization.

In the roll-back scheme, analog and digital simulators simulate their respective portions with their own time steps and maintain roughly the same position in time. When synchronization is needed due to errors in the analog solution, the digital simulation is rolled back to the time of the analog error and re-evaluated. The advantage is that latency can be exploited since analog and digital simulators use their own time steps. However, it may degrade the overall simulation speed, depending on the degree of roll back and frequency with which it is used.

1.6. A SURVEY OF EXISTING SIMULATORS

Mixed-mode simulation has been gaining in popularity over the last few years; as a result, a large number of simulators have been developed. According to the development strategy used, these simulators can be broadly classified into three groups:

1) **Core Modification Approach (C):** The core extension or environment extension approach involves extending an existing analog or digital simulator to cover the levels that are missing. This is accomplished by using existing models or enhancing the models that the simulator can handle, and is typically done in electrical simulators. For example, a simple logic inverter can be implemented in an electrical simulator using controlled sources and other existing primitives to allow "mixed A/D simulation" [SPA88]. Similarly, a logic simulator may be extended to include switches to represent transistors for mixed gate and switch-level simulation. Of course, it would be difficult to extend a logic simulator to incorporate true electrical elements without major modifications to the program.

2) **"Glued" Approach (G):** In this case, two or more existing simulators are combined using either a procedural interface, if the programs are executed in the same address space, or an interprocess communication protocol, if the programs are running in different address spaces. This is an effective solution for companies that have already invested large amounts of time and money maintaining separate simulators and are not willing to abandon them in favor of the development and support of a completely

new simulator. In addition, the input languages for the simulators do not have to be modified and, therefore, have minimal impact on the designer. However, this simple solution also has a number of inherent limitations in terms of efficiency. The processing of bidirectional elements connected across the mixed-mode interface presents a problem, and the time advancement, backup and synchronization of the various simulators that are running concurrently must be addressed. The process of combining a number of different simulators together in this way presents some very difficult implementation and signal mapping problems. Recently, the advent of simulation backplane (**BP**) technology has reduced the barrier of this approach significantly. In this methodology, the various types of simulators are connected through well-defined data-transfer/synchronization mechanisms, collectively referred to as a backplane. While the interfacing issues of this approach are much reduced, the development of the backplane presents an enormous software engineering challenge.

3) **Fully Integrated Approach (I):** This is the most flexible and most efficient approach of the three mentioned here. In this case, the various simulation algorithms are connected via an internal algorithmic backplane and conform to a set of policies defined within the simulator for time-advancement and backup, signal mapping, etc. The algorithms are usually tailored for the mixed-mode environment and are plug-in compatible with the simulator. Therefore, new algorithms may be easily added to this framework. In addition, new models may be added to each of the algorithms defined in the simulator. As a result, there may be several different representations of the same model, commonly referred to as different *views* of the model. The main drawback of this approach is the long development time for this new programming environment, and eventually the support and maintenance associated with it.

Table 1.3 contains a partial list of commercial mixed-mode simulators. iSPLICE3 [SAL89A] is included as a reference for some of the implementation mechanisms described in this book. The summary of the algorithms and techniques used in each level are provided in this table. Note that they vary widely in the simulation modes that are supported and the mixed-mode simulation techniques used. For the mixed-mode timing control, all the time-step synchronization mentioned in Section 1.5 have been used by some commercial program. Some companies use the backplane technology to incorporate tools from other companies into the mixed-mode simulation environment. Therefore, users have the flexibility to choose the best tools for their applications.

The type of mixed-mode interface has not been listed in this table since it is not possible to describe the interface methodology used in each tool in a tabular form. Automatic interface insertion or explicit interface definition are the general approaches but the models may be quite different from simulator to simulator. This issue will be described in depth in Chapter 7. In addition, mixed-mode simulators that combine process, device, and circuit level simulations have been excluded from the table as they are beyond the scope of this book.

1.7. OUTLINE OF THE BOOK

This book focuses on the implementation of fully-integrated mixed-mode simulation and describes event-driven, relaxation-based techniques used in the SPLICE family of programs. While the issues of combining gate, RTL and behavioral levels of simulation are important, they often reduce to simple implementation issues. This book addresses the problem of mixing electrical simulation with gate-level simulation. Since electrical simulation is continuous in nature whereas gate-level simulation is discrete in nature, this particular problem presents a much more interesting chal-

Program/Company	*Type*	*Behavioral/RTL*	*Gate*
Analog Artist Cadence Inc.	G/BP	VHDL, Verilog	D0, D1, Dn, DA, M
Attsim AT&T-Design Automation	I	VHDL, C	D0, D1, Dn, DA, M
ContecSPICE Contec Inc.	C	—	Dn, M
Lsim/Quick II-AccuSim II Mentor Graphic Corp.	G/BP	VHDL, M Language	D0, D1, Dn, DA, M
PSPICE Microsim Corp.	C	PSPICE HDL	D0, D1, Dn, DA, M
Saber Analog Inc.	I	MAST	DA, M
VHDeLDO Anacad Inc.	I	VHDL	Delta Delay
Viewsim/AD Viewlogic System Inc.	G	VHDL, Verilog	D0, D1, Dn, DA, M
iSPLICE3 University of Illinois	I	C	DA, M

Legend D0: zero delay, D1: unit delay, Dn: multiple delay, DA: assigned delay, M: macromodeling, RC: first order RC delay, RCn: piecewise linear RC delay

Table 1.3: Survey of Mixed-Mode Simulators and Their Capabilities

Switch	*Electrical*	*Mixed-Mode Control*
D0, RC, RCn	Direct Matrix	Roll-Back
—	Direct Matrix	Roll-Back
—	Direct Matrix	Analog Control
D0, RC	Direct Matrix	Lock-Step
RC	Direct Matrix	Analog Control
D0, RC, RCn	Direct Matrix	Roll-Back+
—	Mix of Direct and ITA	Lock-Step
—	Direct Matrix	Digital Control
RC, RCn	Mix of Direct and ITA	Roll-Back

+ Calaveras® Scheme Used

Table 1.3 (Continued)

lenge. The book then focuses on the problem of multilevel analog simulation. Initially, the general techniques and algorithms used at each level will be discussed. This is followed by the specific implementation issues associated with an analog multilevel simulation tool called iMACSIM [SIN91].

In Chapter 2, the electrical simulation problem is formulated, and the standard numerical techniques used to solve the problem are presented. Next, the issues associated with the implementation of an efficient time-step control scheme are described. This includes a description of the constraints imposed on the step size by the numerical methods; this is followed by two common time-step control schemes used in circuit simulation programs. In Chapter 3, two properties of waveforms, called *latency* and *multirate behavior*, are defined and used to motivate the need for new simulation methods. Then, the relaxation methods are introduced and their convergence properties are described. First, the linear Gauss-Jacobi (GJ) and Gauss-Seidel (GS) methods are reviewed. Next, the nonlinear relaxation and waveform relaxation methods are described. The requirement for partitioning to improve the convergence speed of relaxation methods is presented at the end of the chapter.

In Chapter 4, a number of algorithms based on nonlinear relaxation methods are described. A technique which combines nonlinear relaxation [ORT70] with *event-driven, selective-trace* [SZY75] to exploit waveform latency is presented. This approach is referred to as Iterated Timing Analysis or ITA [SAL83]. Its name is derived from the original work on "timing" simulation pioneered in the MOTIS program [CHA75]. The details of the implementation of ITA are provided. In addition, a partitioning approach that allows MNA elements to be incorporated into a block time point relaxation framework with guaranteed convergence will be described.

Gate level simulation is addressed in Chapter 5 and switch-level simulation is described in Chapter 6. Chapter 5 begins with a description of the evolution of logic state models and delay modeling. The Elogic technique for switch-level timing simulation and modeling is presented in Chapter 6. In Chapter 7, the modeling problems of mixed-mode interfaces and possible solutions are described. The implementation issues associated with a mixed-mode simulator and an automatic mixed-mode partitioner are described in Chapter 8. Simulation results from a large benchmark circuit is also included to demonstrate the typical performance of mixed-mode simulator.

Analog multilevel simulation is discussed in Chapter 9. First, the key issues of analog multilevel simulation are described. Then the techniques to deal with **s**-domain, **z**-domain and mixed continuous-time/discrete-time simulation are presented. A macromodeling and simulation environment is presented at the end of the chapter.

Chapter 10 provides a summary of the book and directions for future work.

CHAPTER 2

ELECTRICAL SIMULATION TECHNIQUES

The features of circuit or electrical simulation are extremely important in mixed-mode simulation as they determine the overall speedup and efficiency of the simulator. This chapter describes the basic theory and foundations for the electrical simulation techniques. First the circuit equations are formulated in Section 2.1 and the standard techniques are described in Section 2.2. The issues pertaining to time step selection and simulation accuracy are also addressed. The limitations of these techniques for large problems are identified and alternative approaches are described in the next chapter.

2.1. EQUATION FORMULATION

General-purpose circuit simulation programs such as ASTAP [WEE73], SPICE2 [NAG75] and SLATE [YAN80] provide a variety of analysis types including dc analysis, time-domain transient analysis, ac analysis, noise analysis and distortion analysis. By far the most CPU-intensive of these analyses is the time-domain transient analysis. The transient analysis problem involves computing the solution of a system of coupled nonlinear differential-algebraic equations over some interval of time, $[0,\mathbf{T}]$. The most general form for the equations describing the circuit behavior is

$$\mathbf{F}(\dot{\mathbf{x}}(\mathbf{t}), \mathbf{x}(\mathbf{t}), \mathbf{u}(\mathbf{t})) = 0 \quad \mathbf{x}(0)=\mathbf{X} \tag{2.1}$$

where, $\mathbf{x}(\mathbf{t}) \in \mathbb{R}^n$ is the vector of unknowns, and may be a mixture of node voltages, branch currents, capacitive charges or inductive fluxes, $\mathbf{u}(\mathbf{t}) \in \mathbb{R}^r$ is a vector of independent sources, $\mathbf{F}$: $\mathbb{R}^n \times \mathbb{R}^n \times \mathbb{R}^r \rightarrow \mathbb{R}^n$, and the initial conditions, $\mathbf{x}(0)$, are specified by the vector $\mathbf{X}$.

Equations of this form arise as a result of the properties of general electronic circuits. For example, the current through a capacitor is a function of the time derivative of the voltage across the capacitor; therefore, Eq. (2.1) is dependent on $\dot{\mathbf{x}}(\mathbf{t})$. Since many devices have nonlinear relationships between their currents and voltages, $\mathbf{F}$ is also usually nonlinear. And finally, as a circuit is constructed from a collection of sparsely connected elements, $\mathbf{F}$ is a sparse function of the components of $\mathbf{x}$. These circuit properties all have some impact on the numerical techniques used to solve the transient simulation problem and the resulting efficiency with which the solution is obtained.

There are a number of different ways to formulate the circuit equations described by Eq. (2.1). The most popular of these are Nodal Analysis (NA) [DES69], Modified Nodal Analysis (MNA) [HO75] and Sparse Tableau Analysis (STA) [HAC71]. These formulations are all based on the application of Kirchoff's Current Law (KCL), Kirchoff's Voltage Law (KVL) and the branch constitutive equations [DES69]. Nodal Analysis is the simplest of the three approaches. It uses KCL, which requires that the sum of the currents entering each node equals the sum of the currents leaving each node. In a circuit containing $\mathbf{n}$+1 nodes, if KCL is written for every node in the circuit, a system of $\mathbf{n}$ equations is obtained assuming that one node is defined as a reference node. The currents in each equation can be replaced with the branch constitutive relations which are functions of the branch voltages (by assumption in NA), and KVL can be used to replace the branch voltages by node voltages. KVL requires that the sum of the voltages around any loop in a circuit be identically zero. The $\mathbf{n}$ node voltages are the unknown variables in this formulation. Note that it must be possible to represent the element and input source currents in terms of their terminal voltages in order apply Nodal Analysis. This requirement excludes current-controlled devices, floating voltage sources[1] and inductors and, therefore, limits the scope of the NA technique. However, inductors and floating voltage sources can be included in NA by simply reorganizing their branch equations as described in [MCC88, WHI85C]. Since the other current-controlled devices are not frequently used in the simulation of integrated circuits, NA is an adequate formulation technique for most practical circuits.

The formulation used throughout the rest of this book is Nodal Analysis. The NA equations are formulated as follows: First, KCL is applied at each node in a circuit with **n** nodes and **b** branches to produce a matrix equation of the form:

$$\mathbf{A}\,\mathbf{i} = 0 \tag{2.2}$$

where $\mathbf{A} \in \mathbb{R}^{\mathbf{n} \times \mathbf{b}}$ is the reduced incidence matrix with entries of either +1, -1 or 0 and $\mathbf{i} \in \mathbb{R}^{\mathbf{b}}$ is the vector of branch currents in the circuit. Element $\mathbf{a}_{\mathbf{ik}}$ of $\mathbf{A}$ is +1 if a particular branch current, $\mathbf{i}_{\mathbf{k}}$, leaves node **i**, -1 if it enters node **i** and 0 if it is not incident at node **i**. If the set of branch currents is divided into the capacitor currents, $\mathbf{i}_{\mathbf{c}}$, and the currents through the resistive elements, $\mathbf{i}_{\mathbf{r}}$, then Eq. (2.2) can be rewritten as

$$\mathbf{A}_{\mathbf{c}}\mathbf{i}_{\mathbf{c}} = -\,\mathbf{A}_{\mathbf{r}}\mathbf{i}_{\mathbf{r}} \tag{2.3}$$

where $\mathbf{A}=[\mathbf{A}_{\mathbf{c}}\,|\,\mathbf{A}_{\mathbf{r}}]$ and $\mathbf{i}=[\mathbf{i}_{\mathbf{c}}\,|\,\mathbf{i}_{\mathbf{r}}]^{\mathbf{T}}$.

Each of the currents due to the nonlinear resistive elements can be replaced by their branch constitutive relations which are all functions of the branch voltages by assumption. The branch voltages, $\mathbf{v}_{\mathbf{b}}$, can be replaced by the node-to-datum voltages, $\mathbf{v}$, using the relation:

$$\mathbf{A}^{\mathbf{T}}\mathbf{v} = \mathbf{v}_{\mathbf{b}} \tag{2.4}$$

which follows from KVL [CHU75]. Then, the right-hand side of (2.3) can

[1] These are voltage sources with neither terminal connected to the ground node.

be written as

$$\mathbf{A_r i_r} = -\begin{bmatrix} \mathbf{f_1(v)} \\ \mathbf{f_2(v)} \\ \cdot \\ \mathbf{f_n(v)} \end{bmatrix} \tag{2.5}$$

where $\mathbf{f_k(v)}$ is the sum of all the currents through the resistive elements connected to node $\mathbf{k}$ as a function of the node voltages, $\mathbf{v}$.

The left-hand side of Eq. (2.3) represents the capacitor currents. The nonlinear capacitors are often specified in terms of their stored charge, $\mathbf{q}$, a function of the voltage across the capacitor, $\mathbf{v_c}$, as follows:

$$\mathbf{q} = \mathbf{q(v_c)}$$

The current flowing through the capacitor can be obtained by taking the time derivative of charge, which can then be related to the capacitance by applying the chain rule:

$$\mathbf{i_{cap}} = \mathbf{\dot{q}(v_c)} = \frac{\mathbf{dq(v_c)}}{\mathbf{dv_c}} \frac{\mathbf{dv_c}}{\mathbf{dt}} = \mathbf{C(v_c)\dot{v}_c} \tag{2.6}$$

Hence, each of the components of $\mathbf{i_c}$ in Eq. (2.3) can be replaced by $\mathbf{C(v_c)\dot{v}_c}$. If Eq. (2.4) is used to replace the branch voltages by node voltages, then $\mathbf{A_c i_c}$ can be transformed into the following:

$$\mathbf{A_c i_c} = \begin{bmatrix} \mathbf{C_{11}(v)} & \cdot\ \cdot & \mathbf{C_{1n}(v)} \\ \cdot & \cdot\ \cdot & \cdot \\ \cdot & \cdot\ \cdot & \cdot \\ \mathbf{C_{n1}(v)} & \cdot\ \cdot & \mathbf{C_{nn}(v)} \end{bmatrix} \begin{bmatrix} \mathbf{\dot{v}_1} \\ \cdot \\ \cdot \\ \mathbf{\dot{v}_n} \end{bmatrix} \tag{2.7}$$

An important assumption which is sufficient to guarantee convergence of relaxation-based simulation techniques (to be described shortly) is that a two-terminal capacitor exists between each node and the reference node. These are referred to as *grounded capacitors*. This requirement is easily satisfied in real circuits where lumped capacitances are always present between circuit nodes and ground in the form of interconnect capacitance, and also between the terminals of active circuit elements and ground in the form of parasitic capacitances. Each grounded capacitor contributes a term to the diagonal of the capacitance matrix. Therefore, the $\mathbf{C_{ii}}$ elements are non-zero for all $\mathbf{i}$. Note that $\mathbf{C_{ij}}$ is zero only if a capacitor does not exist between nodes $\mathbf{i}$ and $\mathbf{j}$ in the circuit.

By combining Eqs. (2.5) and (2.7), one obtains:

$$\begin{bmatrix} \mathbf{C_{11}(v)} & \cdot\cdot & \mathbf{C_{1n}(v)} \\ \cdot & \cdot\cdot & \cdot \\ \cdot & \cdot\cdot & \cdot \\ \mathbf{C_{n1}(v)} & \cdot\cdot & \mathbf{C_{nn}(v)} \end{bmatrix} \begin{bmatrix} \dot{\mathbf{v}}_1 \\ \cdot \\ \cdot \\ \dot{\mathbf{v}}_n \end{bmatrix} = - \begin{bmatrix} \mathbf{f_1(v)} \\ \cdot \\ \cdot \\ \mathbf{f_n(v)} \end{bmatrix} \tag{2.8}$$

This equation can be written in the compact form:

$$\mathbf{C(v(t),u(t))}\ \dot{\mathbf{v}}(\mathbf{t}) = -\,\mathbf{f(v(t),u(t))}, \quad \mathbf{t} \in [0,\mathbf{T}] \tag{2.9}$$

$$\mathbf{v}(0) = \mathbf{V}$$

where $\mathbf{v(t)} \in \mathbb{R}^\mathbf{n}$ is the vector of node voltages at time $\mathbf{t}$, $\dot{\mathbf{v}}(\mathbf{t}) \in \mathbb{R}^\mathbf{n}$ is the vector of time derivatives of $\mathbf{v(t)}$, $\mathbf{u(t)} \in \mathbb{R}^\mathbf{r}$ is the input vector at time $\mathbf{t}$, $\mathbf{C(x(t),u(t))}$ represents the nodal capacitance matrix, and

$$\mathbf{f(v(t),u(t))} = [\mathbf{f_1(v(t),u(t))}, \cdots, \mathbf{f_n(v(t),u(t))}]^\mathbf{T}$$

where $\mathbf{f_k(v(t),u(t))}$ is the sum of the currents charging the capacitors connected to node $\mathbf{k}$.

Equation (2.9) is a set of coupled first-order nonlinear differential equations that uses voltage as a state variable. This is commonly referred to as the capacitance formulation of the transient analysis problem. Alternatively, charge may be used as a state variable rather than voltage. The proper choice of voltage or charge as the state variable depends on the nature of the capacitors in the circuit. If all capacitances are linear, then either voltage or charge may be used as the state variable. However, in circuits with nonlinear capacitors, such as MOS circuits, charge must be used as the state variable due to considerations of charge conservation. That is, in order to keep the total charge in the system constant during the simulation process, charge must be used as the state variable. Examples of charge conservation problems arising from the use of Eq. (2.9) are given in [WAR78, YAN83, WHI85C].

The charge formulation of the circuit equations in normal form is given by

$$\dot{\mathbf{q}}(\mathbf{t}) = \mathbf{i}(\mathbf{q}(\mathbf{t}))$$

where $\mathbf{q_k(v)}$ is the sum of the charges due to the capacitors connected to node $\mathbf{k}$ and $\mathbf{i_k(q)}$ is the sum of the currents charging the capacitors at node $\mathbf{k}$. This equation can be solved to obtain the node charges as a function of time. However, information about charge is of little interest to the circuit designer, who would prefer to have information about the node voltages from the simulator. Therefore, it is preferable to write the charge formulation as

$$\dot{\mathbf{q}}(\mathbf{t}) = \mathbf{i}(\mathbf{q}(\mathbf{t})) = -\,\mathbf{f}(\mathbf{v}(\mathbf{t}))$$

which is obtained by combining Eq. (2.6) and Eq. (2.9). This assumes that $\mathbf{q}$ is an invertible function of $\mathbf{v}$. The charge formulation, including the input sources, $\mathbf{u(t)}$, is given by

$$\dot{\mathbf{q}}(\mathbf{v}(\mathbf{t}),\mathbf{u}(\mathbf{t})) = -\,\mathbf{f}(\mathbf{v}(\mathbf{t}),\mathbf{u}(\mathbf{t})) \qquad (2.10)$$

Both the formulations given by Eqs. (2.9) and (2.10) will be used throughout this book.

2.2. STANDARD TECHNIQUES FOR TRANSIENT ANALYSIS

Equations (2.9) and (2.10) formulated above for the transient analysis of circuits must be solved using numerical techniques since, in general, it is difficult to obtain closed-form solutions. The first step is to apply a numerical integration method to discretize the time derivative, $\dot{\mathbf{x}}(\mathbf{t})$. An integration method divides the continuous interval of time, $[0,\mathbf{T}]$, into a set of $\mathbf{M}$ discrete time points defined by

$$\mathbf{t}_0 = 0\,, \quad \mathbf{t}_{n+1} = \mathbf{t}_n + \mathbf{h}_n\,, \quad \mathbf{t}_M = \mathbf{T}. \tag{2.11}$$

An algebraic problem is solved at each time point, $\mathbf{t}_{n+1}$, to obtain a sequence approximation to the exact solution. The quantity $\mathbf{h}_n$ is referred to as a time step. The selection of proper time-steps for a given problem is an important issue which is described in detail in Section 2.3. An example of a first-order implicit integration method is the backward-Euler (BE) method. To solve $\dot{\mathbf{x}}(\mathbf{t})=\mathbf{f}(\mathbf{x}(\mathbf{t}))$ using BE, the following expression is used:

$$\mathbf{x}(\mathbf{t}_{n+1}) = \mathbf{x}(\mathbf{t}_n) + \mathbf{h}_n\mathbf{f}(\mathbf{x}(\mathbf{t}_{n+1})) \tag{2.12}$$

This equation is implicit in that $\mathbf{x}(\mathbf{t}_{n+1})$ appears on both sides of the equation.

A numerical integration method converts a set of nonlinear differential equations into a set of nonlinear algebraic equations. These algebraic equations must be solved using some numerical method at each time point. The most commonly used method to solve nonlinear equations is the Newton-Raphson method [ORT70]. To solve a system of nonlinear equations, given by $\mathbf{F}(\mathbf{x})=0$, using the Newton-Raphson method, the following iterative equation is used:

$$\mathbf{J_F}(\mathbf{x}^k)(\mathbf{x}^{k+1} - \mathbf{x}^k) = -\mathbf{F}(\mathbf{x}^k) \tag{2.13}$$

where $\mathbf{J_F}(\mathbf{x})$ is the Jacobian matrix and $\mathbf{k}$ is the iteration counter for the method. Each term in the Jacobian matrix, $\mathbf{g}_{ij}$, is given by

$$\mathbf{g}_{ij} = \frac{\partial \mathbf{F}_i}{\partial \mathbf{x}_j} \tag{2.14}$$

where $\mathbf{F}_i$ is the $\mathbf{i}$th component of $\mathbf{F}$ and $\mathbf{x}_j$ is the $\mathbf{j}$th component of $\mathbf{x}$.

Equation (2.13) is iterated until $\|x^{k+1} - x^k\| < \varepsilon_1$ and $\|F(x^{k+1})\| < \varepsilon_2$. Note tha if the problem is linear, then the Newton method produces the correct solution in one iteration.

The Newton method described above converts the set of coupled nonlinear algebraic equations into a set of coupled linear equations given by $\mathbf{Ax} = \mathbf{b}$, where $\mathbf{x} \in \mathbb{R}^n$, $\mathbf{b} \in \mathbb{R}^n$, $\mathbf{A} \in \mathbb{R}^{n\mathrm{x}n}$ and $\mathbf{A}$ is assumed to be non-singular. The matrix $\mathbf{A}$ is relatively sparse, typically having three elements per row [NEW83]. There are essentially two approaches to solving a sparse linear system. One approach is to use *direct methods* (such as LU decomposition) which attempt to exploit the sparse nature of the matrix during the computation. The implementation of these methods involves carefully chosing a data structure and the use of special pivoting strategies to minimize fillins [KUN86]. A second approach to the sparse linear problem is to use *relaxation methods.* The relaxation process involves decoupling the system of equations and solving each equation separately. An iterative method is applied between the equations until convergence is obtained. In effect, the problem of solving one large system containing **n** variables is converted to the problem of solving **n** subsystems each containing one variable.

The standard approach to circuit simulation is based on direct methods and uses the following steps:

1) MNA is used to formulate a system of coupled nonlinear differential-algebraic equations for the circuit.

2) Implicit numerical integration methods are applied to convert the differential equations into a sequence of algebraic equations, which are nonlinear in general.

3) A damped Newton-Raphson method is used to convert the nonlinear equations into linear equations.

4) Direct sparse-matrix techniques are used to solve the linear equations generated by the Newton-Raphson method.

A simple flow chart of the steps in the standard approach is shown in Fig. 2.1. The details of the implementation of this approach in SPICE2 may be found in [NAG75].

This approach has proven to be very reliable and can be used across a variety of different technologies and element types. The most computationally intensive part of this approach is the Newton-Raphson iteration. It is composed of two phases: the formulation phase and the solution. These two phases, represented by steps 3 and 4 above, are repeated at each time point until convergence is obtained. In the formulation phase, the elements in the circuit are processed by calculating their contribution to the Jacobian matrix and the right-hand side vector in Eq. (2.13) to form the system of linear equations. This is also referred to as the function evaluation (or model evaluation) and load phase, and can be very time-consuming because of the complexity of the equations describing the elements in the circuit. For small to medium sized circuits containing MOS devices, the model evaluation and load times dominate the total CPU-time for the simulation [NEW77].

In the second phase of the Newton iteration, the linear equations generated in the first phase are solved using direct methods such as LU decomposition. While this portion has a negligible contribution to the total run time for small circuits, it can in fact dominate the run time for very large circuits (i.e., greater than 1000 nodes in the circuit for SPICE2) [NEW83], as shown in Fig. 2.2. Therefore, any technique which attempts to reduce overall circuit simulation run times must reduce both the model evaluation time and the linear equation solution time to be effective.

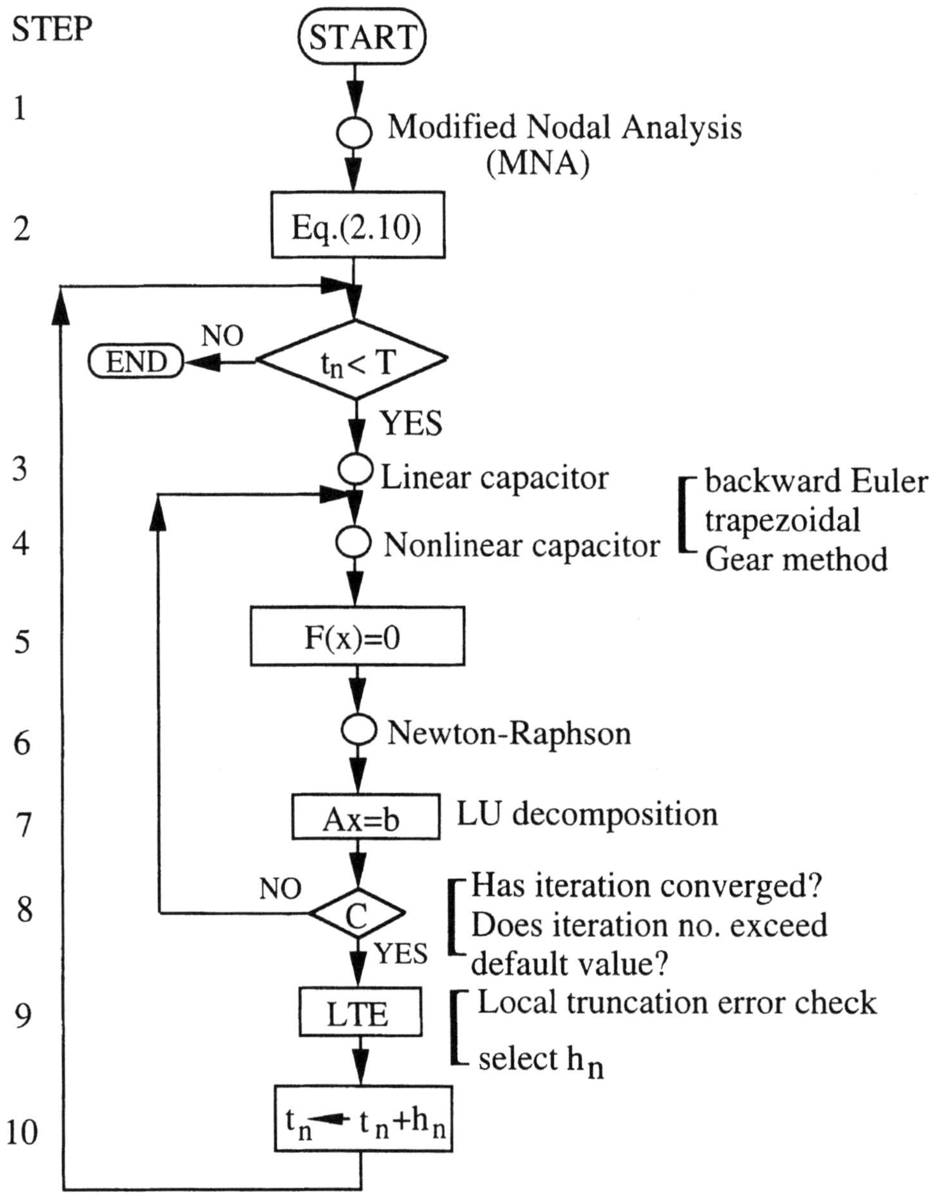

Figure 2.1: SPICE Flow Chart

2.3. TIME-STEP CONTROL: THEORETICAL ISSUES

Time-step control is an important issue in electrical simulation. In this section, the constraints imposed by the numerical techniques on the step sizes used in the integration process are described. Based on these constraints, an efficient time-step control scheme is presented. Ways to further improve the efficiency by using different step sizes to solve different components in the system are presented in the next chapter.

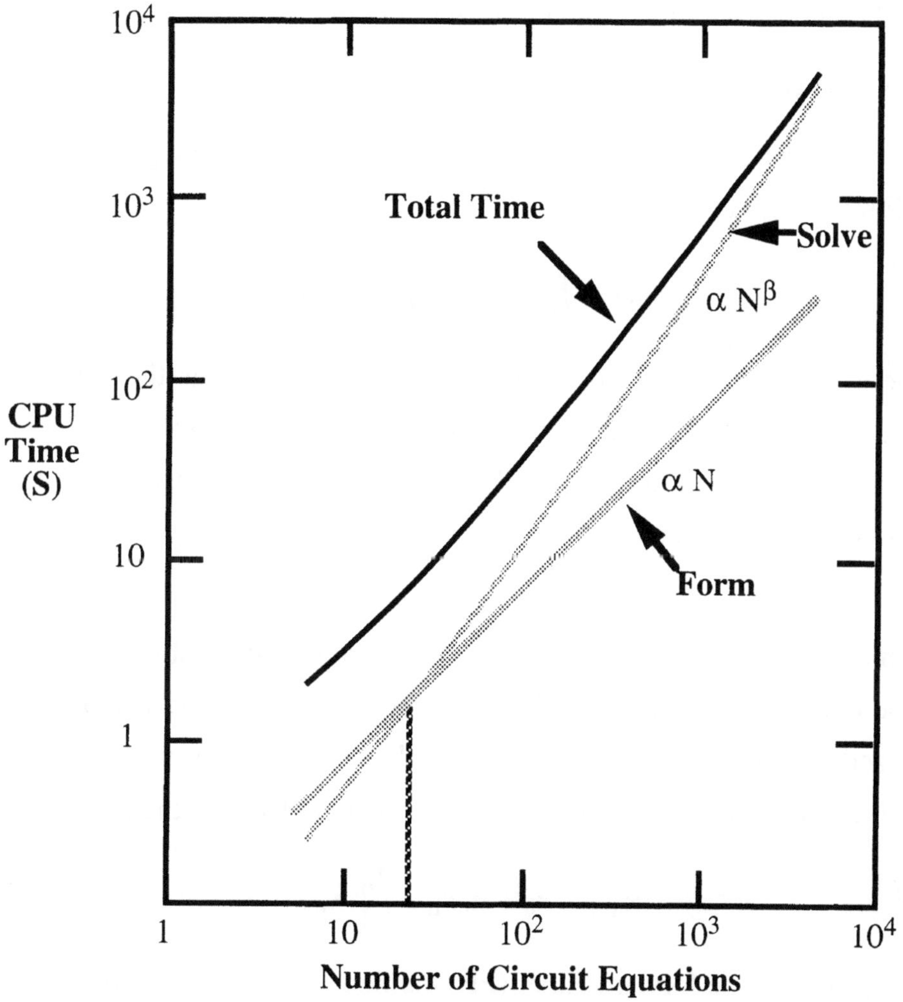

Figure 2.2: CPU-Time vs. Circuit Size in SPICE2

The circuit simulation problem, in its most general form, involves the solution of a system of nonlinear algebraic-differential equations. To simplify the description to follow, the circuit equations are assumed to be a system of differential equations in *normal form*:

$$\dot{\mathbf{x}}(\mathbf{t}) = \mathbf{f}(\mathbf{x}(\mathbf{t}),\mathbf{u}(\mathbf{t}))\,, \quad \mathbf{x}(0) = \mathbf{X}\,, \quad \mathbf{t} \in [0,\mathbf{T}] \tag{2.15}$$

where $\mathbf{u}$ is the set of primary inputs, $\mathbf{x}$ is a vector of unknown circuit variables and $\mathbf{f}$ is some nonlinear function. The vector of values specified as $\mathbf{X}$ are the initial conditions, and the simulation interval is $[0,\mathbf{T}]$.

2.3.1. Constraints on Step Size

The general form of a $\mathbf{k^{th}}$-order linear multistep integration method [GEA71] is given by

$$\mathbf{x}_{n+1} = \sum_{i=0}^{p} \mathbf{a}_i \mathbf{x}_{n-i} + \sum_{i=-1}^{p} \mathbf{h}_{n-i-1} \mathbf{b}_i \dot{\mathbf{x}}_{n-i} \tag{2.16}$$

where $\mathbf{x_n}$ is the computed solution at time $\mathbf{t_n}$, and $\mathbf{h_n}$ is the time-step at time $\mathbf{t_n}$. The $2\mathbf{p}+3$ coefficients, $\mathbf{a_i}$ and $\mathbf{b_i}$, are chosen such that Eq. (2.16) will give the exact solution if the true solution is a polynomial in $\mathbf{t}$ of degree less than or equal to $\mathbf{k}$ [CHU75].

There are two broad classes of integration methods: explicit and implicit[2] [CHU75]. Explicit methods use only the solutions at previous time points to generate the solution at the next time point, and are characterized by $\mathbf{b}_{-1}=0$. A number of explicit integration methods can be derived directly from a Taylor series expansion of $\mathbf{x(t)}$ at the point $\mathbf{t_n}$:

$$\mathbf{x}_{n+1} = \mathbf{x}_n + \mathbf{h}_n \dot{\mathbf{x}}_n + \frac{\mathbf{h}_n^2}{2} \ddot{\mathbf{x}}_n + \cdots \tag{2.17}$$

For example, the forward-Euler (FE) method is obtained by taking the first

[2] Recently, a number of combined integration-relaxation methods used in Timing Simulation [CHA75] have been classified as semi-implicit integration methods [DEM80, NEW83, WHI85C].

two terms of Eq. (2.17):

$$\mathbf{x}_{n+1} = \mathbf{x}_n + \mathbf{h}_n \dot{\mathbf{x}}_n \tag{2.18}$$

This *difference* equation can be formulated in terms of Eq. (2.16) by setting $\mathbf{p}$=0, $\mathbf{a}_0$=1, $\mathbf{b}_0$=1 and all other coefficients to zero. Equation (2.18) implies that each equation can be updated independently, and in parallel, at each time point. For differential equations in the normal form, the solution at each time point can be obtained in one step and does not involve a matrix solution; therefore, the explicit methods are extremely efficient. Unfortunately, these methods are not as useful as implicit methods for circuit simulation because of their stability properties. Implicit methods are characterized by $\mathbf{b}_{-1} \neq 0$ in Eq. (2.16). The backward-Euler (BE) implicit integration method can be derived using a Taylor expansion of $\dot{\mathbf{x}}(\mathbf{t})$ about the point $\mathbf{t}_n$:

$$\dot{\mathbf{x}}_{n+1} = \dot{\mathbf{x}}_n + \mathbf{h}_n \ddot{\mathbf{x}}_n + \frac{\mathbf{h}_n^2}{2} \frac{\mathbf{d}^3 \mathbf{x}_n}{\mathbf{dt}^3} + \cdots \tag{2.19}$$

Using Eq. (2.19) to replace $\dot{\mathbf{x}}_n$ in Eq. (2.17), and ignoring the higher-order terms, the BE scheme is obtained:

$$\mathbf{x}_{n+1} = \mathbf{x}_n + \mathbf{h}_n \dot{\mathbf{x}}_{n+1} \tag{2.20}$$

In this case, $\mathbf{p}$=0, $\mathbf{a}_0$=1, $\mathbf{b}_{-1}$=1 with all other coefficients equal to zero. For nonlinear problems, this implicit equation is usually solved using an iterative method, often requiring a matrix solution. Therefore, the implicit methods are computationally more expensive than explicit methods. The forward-Euler and backward-Euler methods are representative of their respective class of integration algorithms and will be used to illustrate a number of other properties below.

a. Accuracy Constraint

Integration methods provide a numerical approximation to the true solution since, in general, the exact solution of Eq. (2.15) cannot be obtained. The error in the numerical solution is due to a combination of the machine error and the truncation error. The machine error is usually in the form of a round-off error, since finite precision arithmetic is used, and it depends on the floating-point arithmetic unit of the computer being used. The truncation error results from the fact that the Taylor series is truncated after a number of terms and this error depends on the specific integration method. The *local truncation error* (LTE) for general multistep methods is defined as

$$\mathbf{LTE_{n+1}} = \mathbf{x(t_{n+1})} - \mathbf{x_{n+1}} \tag{2.21}$$

where $\mathbf{x(t_{n+1})}$ is the exact solution to Eq. (2.15) at $\mathbf{t_{n+1}}$, and $\mathbf{x_{n+1}}$ is the computed solution obtained from Eq. (2.16). In this definition, it is assumed that $\mathbf{x(t_n)}=\mathbf{x_n}$ and, therefore, only provides information about the error which occurs over a single time-step, hence, its name "local" truncation error. The LTE for the forward-Euler method can be derived using Eq. (2.18):

$$\mathbf{LTE_{n+1}} = \mathbf{x(t_{n+1})} - \mathbf{x_n} - \mathbf{h_n\dot{x}(t_n)} \tag{2.22}$$

Using a Taylor expansion for the first term about $\mathbf{t_n}$, the LTE is given by the first remainder term of the resulting expression:

$$\mathbf{LTE_{n+1}} = \frac{\mathbf{h_n^2}}{2}\mathbf{\ddot{x}(\xi)} \qquad \mathbf{t_n} \le \xi \le \mathbf{t_{n+1}} \tag{2.23}$$

If $\mathbf{E_A}$ is some user allowable error tolerance for the problem, the accuracy constraint is

$$\frac{h_n^2}{2}\ddot{x}(\xi) \leq E_A \qquad t_n \leq \xi \leq t_{n+1} \tag{2.24}$$

This presents a bound on the step size which is given by

$$h_n \leq \sqrt{2E_A/\ddot{x}(\xi)} \tag{2.25}$$

If this constraint is not satisfied, the solution must be rejected and a new solution is computed using a smaller step size. Since the exact value of ξ is not known, the LTE is usually estimated using techniques to be described in Section 2.4.1.

The backward-Euler method has an LTE given by

$$LTE_{n+1} = x(t_{n+1}) - x_n - h_n\dot{x}(t_{n+1}) \tag{2.26}$$

By expanding $x(t_n)$ in a Taylor series about t_{n+1} and applying the results to Eq. (2.26), the LTE is obtained by retaining the first remainder term:

$$LTE_{n+1} = -\frac{h_n^2}{2}\ddot{x}(\xi) \qquad t_n \leq \xi \leq t_{n+1} \tag{2.27}$$

Note that the error made in one step is $O(h^2)$ in both the FE and BE methods; hence, the accuracy bound on the step size is similar in both cases. However, the behavior of the global error, due to the accumulation of the local errors, may be quite different for the two methods and this difference strongly recommends the use of one method over the other. This characteristic is associated with the stability of the integration method.

b. Stability Constraint

The general stability characteristics of numerical integration methods applied to nonlinear differential equations are difficult to obtain. Usually the results are inferred from the analysis of a simple linear test problem [GEA71]:

$$\dot{\mathbf{x}}(\mathbf{t}) = -\lambda\mathbf{x}(\mathbf{t}) \ , \quad \mathbf{x}(0) = \mathbf{x}_0 \tag{2.28}$$

for which the solution is known to be

$$\mathbf{x}(\mathbf{t}) = \mathbf{x}_0\mathbf{e}^{-\lambda\mathbf{t}} \tag{2.29}$$

and, in general, λ is complex. This linear problem is useful because it is easy to analyze and provides information about the local behavior of non-linear problems (i.e., when the step size is small). The problem is usually analyzed with $\mathbf{Re}(\lambda)>0$ so that the solution to Eq. (2.28) is stable. To further simplify the analysis, a fixed time-step is assumed. For example, if the FE method is used to solve Eq. (2.28), the following difference equation is obtained:

$$\mathbf{x}_{n+1} = \mathbf{x}_\mathbf{n} - \lambda\mathbf{h}\mathbf{x}_\mathbf{n} = \mathbf{x}_\mathbf{n} - \sigma\,\mathbf{x}_\mathbf{n}$$

where $\sigma = \lambda\mathbf{h}$. Therefore,

$$\mathbf{x}_{n+1} = (1-\sigma)\mathbf{x}_\mathbf{n} = (1-\sigma)^{\mathbf{n}+1}\mathbf{x}_0$$

The region of absolute stability is defined as the set of all complex values of σ such that $\mathbf{x}_{n+1}$ remains bounded as $\mathbf{n}\to\infty$. For FE, it consists of all σ such that

$$|1-\sigma|\leq 1 \tag{2.30}$$

which produces the following constraint for real values of λ:

$$0 \leq \sigma \leq 2.$$

Therefore the time-step must lie in the range:

$$0 \leq \mathbf{h} \leq \frac{2}{\lambda}. \tag{2.31}$$

If step sizes outside this range are used, the computed solution will become unstable even though the true solution is stable. For BE, the difference equation is

$$\mathbf{x_{n+1} = x_n - \sigma\, x_{n+1}}$$

Hence,

$$\mathbf{x_{n+1}} = \frac{1}{(1+\sigma)^{n+1}} \mathbf{x_0}$$

which results in the following requirement for stability:

$$\frac{1}{|1+\sigma|} \leq 1 \tag{2.32}$$

Considering only real values of $\boldsymbol{\lambda}$, the method produces a stable solution for all $\mathbf{h}\geq 0$. Ideally, an integration method should produce a stable solution if the true solution is stable for any step size; this is the case for the BE method but not for FE. This property highly recommends the use of the BE method over the FE method since the step size can be selected based on accuracy considerations alone. For the general case when $\boldsymbol{\lambda}$ is complex, the region of Absolute stability for the BE integration method includes the entire right-half σ-plane. An integration method with this property is said to be A-stable [CHU75].

The forward-Euler and backward-Euler methods are examples of first-order integration methods. Higher-order methods with smaller local truncation errors can be constructed by taking more terms in the Taylor expansions of Eqs. (2.17) and (2.19). Integration methods with small LTEs are preferred as they allow larger time-steps to be used. For example, the trapezoidal method is a second-order integration method given by

$$\mathbf{x_{n+1} = x_n + \frac{h_n}{2}(\dot{x}_{n+1} + \dot{x}_n)} \tag{2.33}$$

and is quite popular as it is the most accurate A-stable method [CHU75]. The LTE for the trapezoidal method can be shown to be [CHU75]:

$$\mathbf{LTE}_{n+1} = -\frac{\mathbf{h}_n^3}{12}\frac{\mathbf{d}^3\mathbf{x}}{\mathbf{dt}^3}(\xi) \qquad \mathbf{t}_n \le \xi \le \mathbf{t}_{n+1} \tag{2.34}$$

Since the error is $\mathbf{O}(\mathbf{h}^3)$, it is often the case that a much larger step size can be used, compared to the BE method, for a given value of $\mathbf{E}_\mathbf{A}$.

c. Stiff-Stability Constraint

Another consideration in the choice of integration methods is the issue of stiffness. A stiff problem is one that exhibits time-scale variations of several orders of magnitude in the solution. A simple example of stiffness is the case of a fast initial "transient" in the solution, which dies quickly, followed by a slower "steady-state" solution. To handle this type of behavior, it is natural to use small time steps in the transient portion to accurately follow the solution and then to increase the step size for the remainder of the solution. However, this strategy may lead to instability of the integration method, especially for explicit integration methods. For example, if the test problem in Eq. (2.28) is solved using FE in the interval $[0,10^6\tau]$, where $\tau = 1/\lambda$, and $\lambda \in \mathbb{R}$, the time-step constraint given in Eq. (2.31) would be imposed in the entire interval even though the solution decays to zero in approximately 5τ. If the step size is increased beyond this stability bound, the solution will become unstable. On the other hand, if the size is kept within the constraint imposed by stability, the number of time points would be very large.

There are other situations which feature this kind of time-scale variation. A stiff problem is generated if the interval of time over which the system is integrated is large compared to the smallest time constant in the circuit, or if the circuit time constants themselves are widely separated. In addition, if the rise or fall time of an input waveform is widely separated from the circuit time constants, the problem also is considered to be stiff.

Integration methods which are appropriate for solving stiff problems should have regions of Absolute Stability which cover most of the right-half complex σ-plane so that the time-step can be selected based on the accuracy considerations alone. Explicit methods are not well-suited to stiff problems since their regions of Absolute Stability are usually very small. The A-stable integration methods are well-suited to stiff problems, but other implicit methods (for example, see [CHU75]) may be prone to instability when solving stiff problems. Gear proposed a family of integration methods called *stiffly-stable* methods [GEA71] which have the form:

$$\dot{\mathbf{x}}_{n+1} = \frac{1}{\mathbf{h}_n} \sum_{i=0}^{\mathbf{k}} \alpha_i \mathbf{x}_{n+1-i} \tag{2.35}$$

The values for α_i are chosen such that a **k**th-order method is exact if the true solution is a **k**th-order polynomial. The methods of order k=1 and k=2 are both A-stable algorithms. The methods of order k=3 up to k=6 are not A-stable, but they do have stability regions which are quite suitable for the integration of stiff problems [GEA71]. These methods are also referred to as Backward-Differentiation Formulas (BDF) [BRA72]. A variable-order method, also proposed by Gear [GEA71], uses the integration order which allows the largest step size at each time point. This technique was implemented in the SPICE2 program [NAG75] and it was found that, even though the order could be varied from k=1 up to k=6, a second-order method was used most often in the computation. The reason for this was attributed to the nature of the nonlinearities in the circuit simulation problem (described in the next section) and nature of the solution waveforms. Therefore, most circuit simulators use a low-order implicit integration method with guaranteed stability properties so that the step sizes can be selected based on accuracy considerations alone.

2.3.2. Solution of Nonlinear Equations

When solving linear dynamic circuits, the accuracy and stability requirements of the numerical integration method are the only constraints on the step size used. Furthermore, linear problems can be solved in one "iteration" (i.e., one matrix solution) at each time point. Therefore, the amount of computation is directly proportional to the number of time points used. This is not true for nonlinear dynamic circuits, assuming that an implicit integration method is used. In fact, the cost of computing a solution at each time point is a function of the number of iterations used to solve the nonlinear algebraic problem. Consider the differential equation

$$\dot{\mathbf{x}}(\mathbf{t}) = \mathbf{f}(\mathbf{x}(\mathbf{t})) \tag{2.36}$$

where $\mathbf{f}(\mathbf{x})$ is some nonlinear function. If the BE method is used to solve Eq. (2.36), the following equation is obtained:

$$\mathbf{x}_{n+1} = \mathbf{x}_n + \mathbf{hf}(\mathbf{x}_{n+1}) = \mathbf{G}(\mathbf{x}_{n+1}) \tag{2.37}$$

This nonlinear algebraic equation can be solved using a variety of techniques including fixed-point iteration and Newton's method. The approach usually taken in circuit simulators is to use Newton's method or one of its variants. Rewriting Eq. (2.37) as

$$\mathbf{F}(\mathbf{x}_{n+1}) = \mathbf{x}_{n+1} - \mathbf{x}_n - \mathbf{hf}(\mathbf{x}_{n+1}) = 0 \tag{2.38}$$

the Newton-Raphson method to solve this equation is given by the expression [ORT70]:

$$\mathbf{x}_{n+1}^{k+1} = \mathbf{x}_{n+1}^{k} - \mathbf{F}(\mathbf{x}_{n+1}^{k})/\mathbf{F}'(\mathbf{x}_{n+1}^{k}) \tag{2.39}$$

where $\mathbf{k}$ is the Newton iteration counter. In circuit terms, the Newton method replaces each nonlinear device in the circuit by a linearized model based on operating point information. This process converts the nonlinear circuit into a linear equivalent network. The linearized network is solved using standard linear circuit analysis techniques [CHU75]. The Newton method involves repeating the above steps until convergence is obtained.

To guarantee convergence of the Newton method, the functions $\mathbf{F(x)}$ and $\mathbf{F'(x)}$ must be continuous in an open neighborhood about $\mathbf{x^*}$, $\mathbf{F'(x^*)} \neq 0$, and the initial guess, $\mathbf{x}^0$, must be close to the final solution. The Newton method is preferred over the simpler fixed-point method for several reasons. The main reason is that the fixed-point algorithm is not well-suited to stiff problems. It also imposes a bound on the time-step to guarantee convergence. Another reason is due to the quadratic convergence property of the Newton method. That is, if, in addition to the above conditions, $\mathbf{F''(x^*)}$ exists, then for some $\mathbf{k>K}$ the difference between successive iterations and the true solution satisfies the relation [ORT70]:

$$|\mathbf{x}^{k+1} - \mathbf{x}^*| \leq \mathbf{c}\,|\mathbf{x}^k - \mathbf{x}^*|^2$$

In practice, this quadratic convergence behavior occurs close to the final solution. Hence, it is important to provide an initial guess which is close to the final solution. In general, it is difficult to provide a reasonable starting guess for the Newton method. However, for the transient analysis problem it is possible to generate a good initial guess, especially if a capacitor exists between each node and the ground node[3]. For example, the solution at the previous time point is a good starting guess for the Newton method at $\mathbf{t}_{n+1}$. A better approach is to use an explicit integration method [BRA72]:

$$\mathbf{x}_{n+1}^0 = \sum_{i=1}^{k+1} \gamma_i \mathbf{x}_{n+1-i} \tag{2.40}$$

where the γ_i values are obtained by requiring that the predictor, $\mathbf{x}_{n+1}^0$, be correct if the solution is a $\mathbf{k}$th-order polynomial. Usually a $\mathbf{k}$th-order predictor is used with a $\mathbf{k}$th-order integration method.

---

[3] A capacitor to ground at each node implies some smoothness in the solution since it prevents instantaneous changes in the voltage at the node. Therefore, as $\mathbf{h} \rightarrow 0$, $\mathbf{x}_{n+1} \rightarrow \mathbf{x}_n$.

The time-step also has some influence on the convergence speed of the Newton method. An intuitive reason for this can be given in circuit terms: the Newton method converts a nonlinear circuit into an associated linear circuit, as mentioned previously. As the step size is made smaller, the values of linearized circuit elements begin to approach their values at the previous time point. Therefore, the circuit will behave almost linearly in this interval and convergence can be obtained in very few iterations, possibly even a single iteration. On the other hand, if the step size is too large, a good starting guess may be difficult to generate, and could lead to either slow convergence or nonconvergence. If nonconvergence should occur, the time-step must be rejected and a smaller step used in its place. Hence, in some cases, it may actually be more efficient to use two small steps rather than one large step.

2.4. TIME-STEP CONTROL: IMPLEMENTATION ISSUES

The simplest time-step selection scheme is to use the same time-step throughout the interval of interest, [0,**T**]. That is, use a *fixed* time-step. Unfortunately, there are a number of constraints on the step size which may require that **h** be extremely small, resulting in a large number of time points. These constraints arise from the accuracy, stability and stiff-stability properties of a numerical integration method. For a fixed-step approach, the step size would have be chosen such that it satisfies these constraints under worst-case conditions. A better approach is to vary the step size during the simulation in accordance with the variation in the constraints. For a given problem, the allowable step sizes depend primarily on the properties of the specific integration method being used. In this section, the main considerations in the implementation of an efficient time-step control for circuit simulation are described. It includes a discussion of LTE time-step control, iteration count time-step control and the effect of input sources on time-step selection.

2.4.1. LTE Time-Step Control

In LTE time-step control, the user provides two accuracy control parameters, ε_a and ε_r, which are the absolute and relative errors permissible in each integration step. They are combined to form a user error tolerance:

$$E_{UserLTE} = \varepsilon_a + \varepsilon_r \times \max |x_{n+1}, x_n|$$

The general form of the local truncation error for most multistep integration methods of order **k** is given by [GEA71,CHU75]

$$LTE_{n+1} = \tilde{C}_k h^{k+1} x^{(k+1)}(\xi) \qquad t_n \le \xi \le t_{n+1} \tag{2.41}$$

where $\tilde{C}_k$ is a constant which depends on the coefficients of Eq. (2.16) and the order of the method. Since the value of $x^{(k+1)}(\xi)$ is not known, in general, it must be estimated in some way using the numerical solutions. Typically a divided-difference approximation is used. The first divided-difference is defined as

$$DD_1(t_{n+1}) = \frac{x_{n+1} - x_n}{h_n}$$

and the **k**+1st divided-difference is defined as

$$DD_{k+1}(t_{n+1}) = \frac{DD_k(t_{n+1}) - DD_k(t_n)}{\sum_{i=0}^{k-1} h_{n-i}}$$

Then the estimate for the derivative term in Eq. (2.41) is (see [NAG75] for derivation)

$$x^{(k+1)}(\xi) \approx (k+1)!\, DD_{k+1}(t_{n+1}).$$

The LTE estimate is then

$$E_k = C_k h^{k+1} DD_{k+1}(t_{n+1})$$

For the BDF integration methods [BRA72], the LTE can be estimated in a more convenient way. The estimate is the calculated using difference between the computed solution x_{n+1} and the predicted value $x^P(t_{n+1})$. For a **k**th-order BDF method, the following expression is used:

$$E_k = \left[\frac{h_n}{t_{n+1} - t_{n-k}} \right] (x_{n+1} - x^P(t_{n+1}))$$

The expression for $x^P(t_{n+1})$ is given in Eq. (2.40). The computed solution x_{n+1} is accepted if

$$|E_k| < E_{UserLTE} \tag{2.42}$$

One way of implementing this check is to take the ratio of the allowable LTE and the actual LTE:

$$r = \frac{|E_{UserLTE}|}{|E_k|} = \frac{|C_k h_{max}^{k+1} \ddot{x}(\xi)|}{|C_k h_n^{k+1} \ddot{x}(\xi)|}$$

Noting that both errors are $O(h^{k+1})$, it follows that

$$r = \left(\frac{h_{max}}{h_n} \right)^{k+1}$$

and

$$r_{LTE} = \frac{h_{max}}{h_n} = (r)^{\left(\frac{1}{k+1}\right)}$$

The comparison test given in Eq. (2.42) becomes

$$r_{LTE} > 1.0$$

to accept the computed solution. The advantage of this ratio is that it can also be used to select the step size for the next integration step. Therefore, the next recommended step size is given by

$$h_{n+1} = r_{LTE} h_n \tag{2.43}$$

In practice, Eq. (2.43) may occasionally recommend rather abrupt changes in the step size. A number of experiments have shown that rapid changes in step size may introduce stability problems [BRA72]. Intuitively, the step sizes should follow the smoothness of the solution. To ensure that the changes in the step size are indeed gradual, it is convenient to set upper and lower bounds on the changes in step size. In fact, three regions can be defined as follows:

- if $\mathbf{r_{LTE}} < 1.0$, reduce the step size by MAX($\mathbf{s_l}$, $\mathbf{r_{LTE}}$)
- if $1.0 \leq \mathbf{r_{LTE}} < \alpha$, maintain the same step size
- if $\mathbf{r_{LTE}} \geq \alpha$, increase the step size MIN($\mathbf{s_u}$, $\beta\, \mathbf{r_{LTE}}$)

The time step may be reduced at most by the factor $\mathbf{s_l}$ and increased at most by the factor $\mathbf{s_u}$. The α factor permits the same step size to be used a number of times. Typically, α=1.2, $\mathbf{s_l}$=0.25 and $\mathbf{s_u}$=2.0. Note that a multiplying factor β has also been introduced as part of the growth factor. The β factor is a way of making the time step selection somewhat conservative. Since the LTE can only be estimated, it may occasionally be optimistic [YAN80]. If so, the time step would be rejected and a smaller step used unnecessarily. The β factor reduces the likelihood of this happening and a typical value is 0.9.

2.4.2. Iteration Count Time-Step Control

As mentioned before, the use of large steps is not necessarily the most efficient approach for nonlinear circuits, especially if relaxation is used. In fact, if the time step is too large, the iterative method may not converge, which would force the time step to be rejected, resulting in wasted effort. This suggests that the time step control should also be controlled by the nonlinearity of the problem.

A number of programs use a time step control based on nonlinearity considerations alone (e.g., SPICE2, ADVICE [NAG80], MOTIS3) called *iteration count* time step control. This strategy minimizes the total number of Newton iterations used during the simulation. The step sizes are selected as follows. If the number of iterations is larger than $\mathbf{N_{high}}$, the step size is reduced by some factor. If the number of iterations is less than $\mathbf{N_{low}}$, the step size is increased by some factor. Otherwise, the step size remains the same. The idea is to use approximately the same number of iterations at each time point.

While this strategy is certainly effective at reducing the overall computation time, it is prone to accuracy problems [NAG75]. For example, for linear circuits the step size would always be increased since the solution is always obtained in one "iteration" at each time point. For weakly nonlinear circuits, the same sort of effect would be observed. Therefore, this approach, when used by itself, is not recommended since it does not control the numerical integration errors directly. However, the iteration count time-step control can be used in conjunction with the LTE-based time step control. In this case, if too many iterations were required to converge, a somewhat smaller step size could be used in the next integration step. If too few iterations are used, a slightly larger step size can be used. The method could be implemented by making the growth factor dependent on the number of iterations used to compute the solution. Of course, if convergence is not obtained in a specified number of iterations, the time step should be rejected and a smaller step used in its place.

CHAPTER 3

RELAXATION-BASED SIMULATION TECHNIQUES

The overall goal in circuit simulation is to generate the solution as efficiently as possible while providing the desired level of accuracy. As described in the last chapter, the standard approach to solving Eq. (2.1) is to use a numerical integration method. One way to make the integration process efficient is to simply minimize the total number of time points used. That is, at any stage during the simulation, take the largest step possible that provides the required accuracy. This strategy is effective for linear problems, assuming that the numerical integration method has guaranteed stability properties but does not guarantee a smaller runtime. In fact, for nonlinear problems, it may be more efficient to take smaller steps so that the iterative method used to solve the nonlinear algebraic equations converges in fewer iterations. Using small time-steps also improves the accuracy of the solution. Therefore, minimizing the total number of iterations is a more useful goal in reducing the amount of computation.

The cost of each iteration is proportional to the number of model evaluations[1] performed, assuming that the linear equation solution time is small. Therefore, the number of model evaluations used in the solution process is a good measure of the amount of computation. Based on this argument, the objective for the efficient solution of the differential equations in Eq. (2.1) should be to minimize the total number of model evaluations.

---

[1] A model evaluation usually refers to the calculation of the currents and conductances for a MOS or bipolar transistor, or some equivalent amount of computation.

A number of researchers have attempted to reduce the computation time for expensive model evaluations by using lookup tables for active devices [CHA75, SHI82, SHI83, BUR83, GYU85, SAK85B]. In this approach, a number of tables of device characteristics are generated prior to the analysis, and simple table lookup operations are performed during the analysis in place of the expensive analytic evaluations. Points which are not available in the tables are interpolated using polynomial interpolation or splines. One drawback of this approach is that there may be a substantial memory requirement for these tables, depending on the level of accuracy desired, but it is usually justified by the improvement in computation speed. Current research in this area involves reducing the memory requirements without sacrificing either the computational advantage or the accuracy of the device models. Further details on this topic may be found in the references listed above.

In this chapter, the focus is on reducing the total number of expensive model evaluations by minimizing the number of time points computed for each waveform. This is accomplished by using relaxation-based techniques to exploit the waveform properties such as latency and multirate behavior. Section 3.1 begins by introducing the general concepts of waveform latency and multirate behavior. In Section 3.2, the various relaxation-based techniques that are used to exploit latency and multirate behavior are examined. In Section 3.2.4, the circuit partitioning issues for relaxation methods are addressed.

3.1. LATENCY AND MULTIRATE BEHAVIOR

Most circuit simulators employing direct methods use a single common time step for the whole system and, hence, compute the solution of every variable at every time point. The time-step at each point is based on the fastest changing variable in the system, i.e., the **n**+1st time point is given by

$$\mathbf{t}_{n+1} = \mathbf{t}_n + \mathbf{h}_n$$

where $\mathbf{h}_n$ is the integration step size determined by

$$\mathbf{h}_n = \min(\mathbf{h}_{1,n}, \mathbf{h}_{2,n}, \cdots, \mathbf{h}_{N,n})$$

and $\mathbf{h}_{i,n}$ is the recommended step size for with the **i**th variable at $\mathbf{t}_n$. As a result, many variables are solved using time steps which are much smaller than necessary to compute their solutions accurately. For example, the computed points of a waveform from a large digital circuit, simulated using direct methods, are shown in Fig. 3.1(a). Note that there are many more points than necessary to represent the waveform accurately, especially in the regions when the waveform is not changing at all. The extra points are due to some other variable changing rapidly in the same region of time. The same waveform is shown Fig. 3.1(b) with only the minimum number of points necessary to represent it accurately.

Since the objective in circuit simulation is to provide an accurate solution while minimizing the number of expensive model evaluations, one way to achieve this goal is to reduce the number of time points computed for each waveform. A number of circuit simulators have attempted to improve the efficiency in this manner by exploiting a property of waveforms called *latency* [NAG75, NEW78, RAB79, YAN80, SAK81]. While the general concept of latency includes any situation where the value of a variable at a particular time point can be computed accurately using some explicit formula, it usually refers to the situation where a variable is not changing in time and its solution can be obtained from the explicit equation:

$$\mathbf{x}_{n+1} = \mathbf{x}_n \tag{3.1}$$

That is, the value $\mathbf{x}_{n+1}$ is not computed using a numerical integration

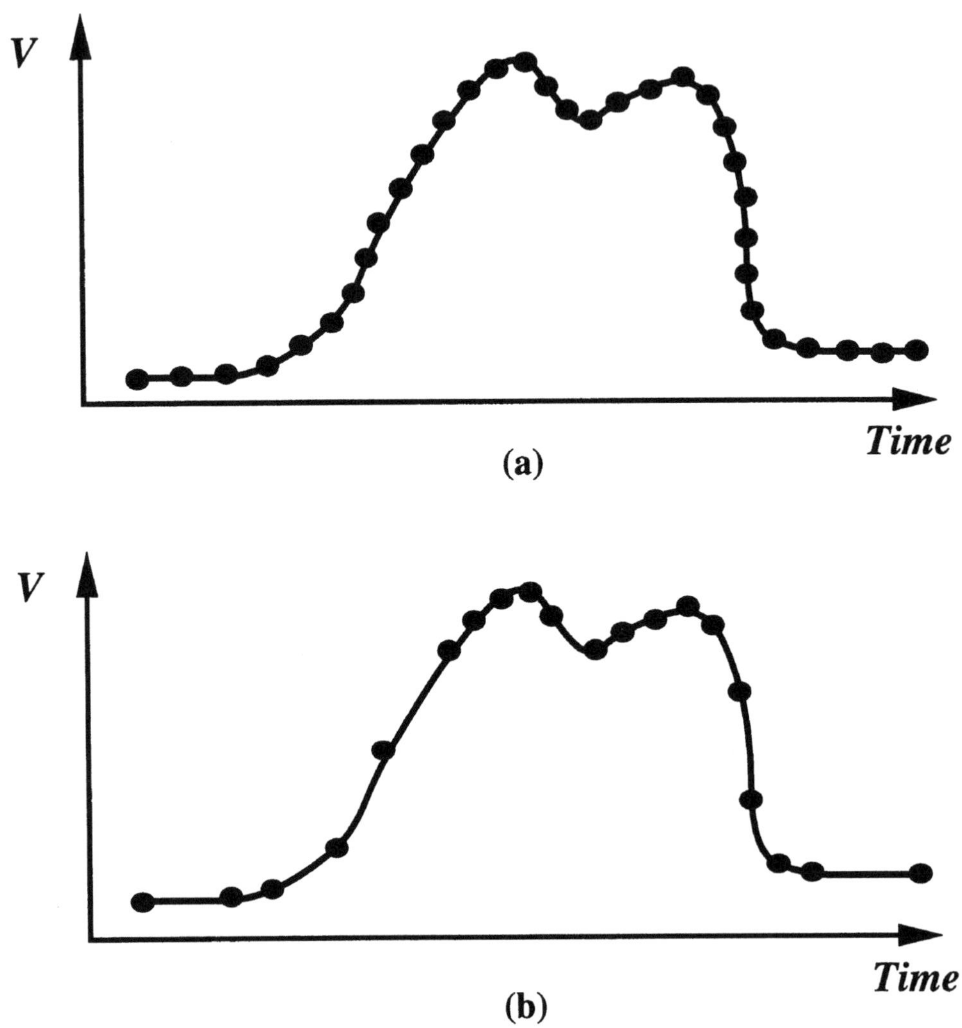

Figure 3.1: Effect of Solution by Direct Methods

formula but instead is simply updated using the value at the previous time point. For example, the waveform shown in Fig. 3.2(a) has three latent periods, and ideally the value of **x** does not need to be computed in any of these regions.

In the SPICE program [NAG75], latency exploitation is performed using a *bypass* scheme. In this technique, each device is checked to see if any of its associated currents and node voltages have changed significantly since the last iteration. If not, the same device conductances and current are also used in the next iteration. However, the checking operation is somewhat expensive, especially if the circuit is large and most of the devices are latent. In general, latency exploitation involves the use of a model describing the behavior of a particular variable as a function of time over a given interval. The simple model described in Eq. (3.1) can be considered as a "zeroth-order" latency model. Higher-order latency models can be constructed if the solution is known to have a specific form (i.e., polynomial, exponential) or if the solution for the variable can be obtained in closed form. For example, a first-order latency model given by

$$\mathbf{x}_{n+1} = \mathbf{x}_n + \mathbf{h}_n \frac{\mathbf{I}}{\mathbf{C}}$$

can be used in the case of an ideal current source, with current **I**, charging a linear capacitor, **C**. Usually a latency model can only be used over a portion of the simulation interval. Therefore, the validity of the model must be monitored and its use must be discontinued when the model is thought to be invalid. The latency model used in this context has also been called a *dormant* model [SAK81].

In practice, only the zeroth-order form of latency can be exploited easily since the higher-order forms are difficult to construct for general nonlinear circuits. To exploit this simple form of latency, some mechanism is necessary to detect that the signal value is not changing appreciably. The waveform is considered to be latent at that point, and its associated variable is updated using Eq. (3.1) at subsequent time points. A second mechanism is used to determine when the latency model is invalid, and from that point onward the variable is computed in the usual way. Hence, the waveform is only computed at time points when the signal is changing. Event-driven, selective trace can be used to exploit latency, as described in the next chapter, without incurring the overhead of the bypass scheme.

It is only useful to exploit this simple form of latency when some variables in the circuit are changing while other variables are stationary, since direct methods can adequately handle the case when all variables are active or latent. In fact, the "useful" form of zeroth-order latency can be viewed as a subset of a more general property of waveforms called *multirate behavior* which is illustrated in Fig. 3.2(b). Multirate behavior refers to signals changing at different rates, relative to one another, over a given interval of time. MOS circuits inherently exhibit this kind of behavior because of different transistor sizes and different capacitance values at each node. Exploiting this general property can reduce significantly the number of time points computed for each waveform since large steps can be used for variables changing very slowly while smaller steps can be used for rapidly changing variables.

The basic strategy to speed up circuit simulators suggested above is to take advantage of the relative inactivity of large circuits by reducing the number of time points computed. However, the actual speed improvement obtained by solving the equations in this manner depends on the two main factors:

1) The "amount" of latency and multirate behavior exhibited by the circuit during the simulation, and

2) The efficiency of techniques used to exploit the two properties.

The first point refers to the maximum speed improvement that can be obtained if the two waveform properties are exploited fully, and this factor depends on the circuit size and the activity in the circuit generated by the external inputs. The second factor depends on the actual number of points computed and the work required to compute each point.

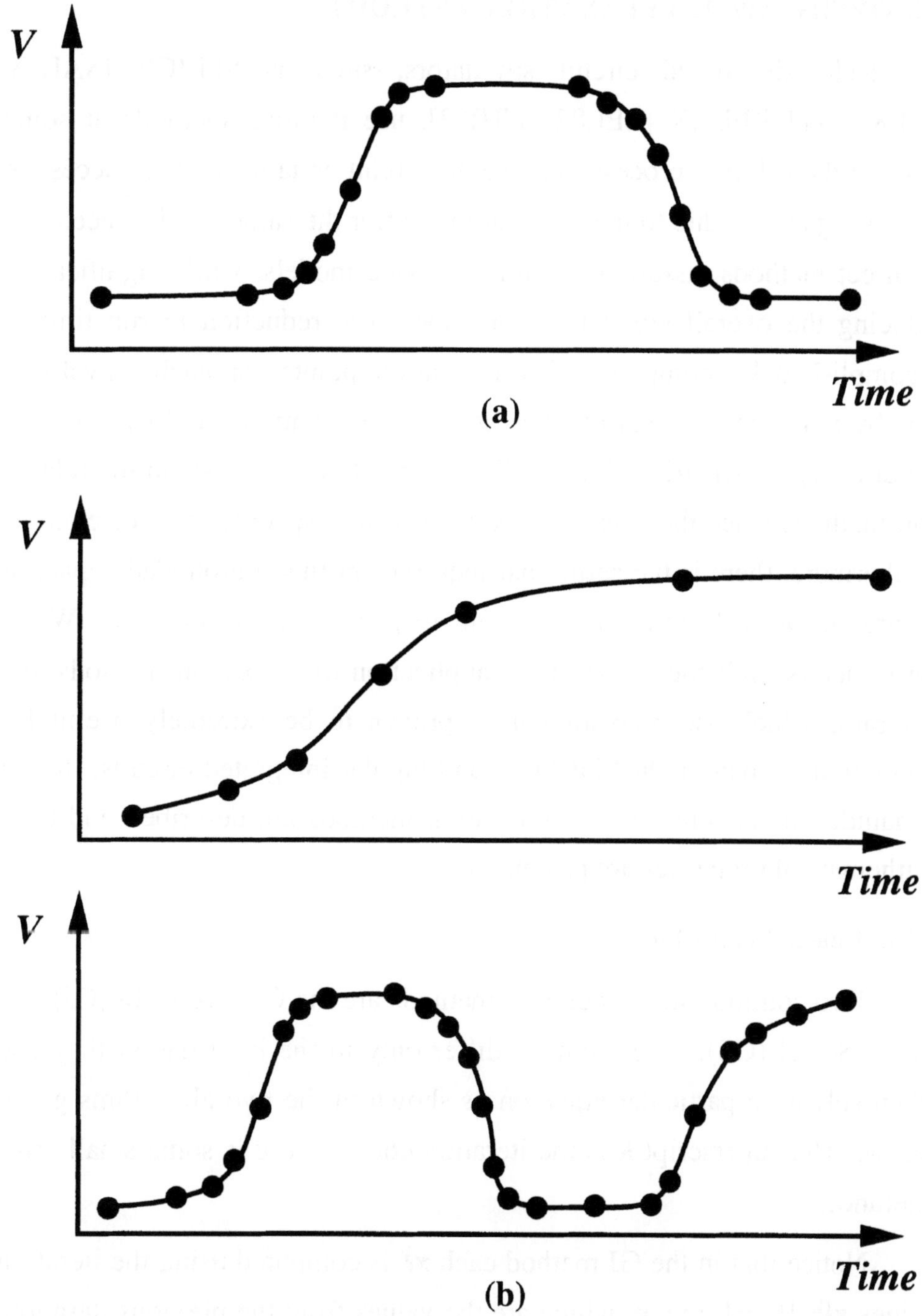

Figure 3.2: Waveform Properties (a) Latency (b) Multirate Behavior

3.2. OVERVIEW OF RELAXATION METHODS

Relaxation-based circuit simulators, such as SPLICE [SAL83, KLE84] and RELAX [LEL82, WHI83], use iterative methods at some stage of the solution process to solve the circuit equations. The success of these programs is due to the fact that they offer the same level of accuracy as direct methods, assuming identical device models, while significantly reducing the overall simulation run time. The reduction in run time is accomplished by computing fewer solution points for each waveform, thereby reducing the total number of model evaluations, and by avoiding the direct sparse-matrix solution. However, a tradeoff exists in the relaxation methods since they can only be applied to a specific class of circuits. Furthermore, there is the additional requirement that a grounded capacitor be present at each node in the circuit to guarantee convergence. While these factors limit the scope of the application of relaxation methods, the programs which use relaxation have proven to be extremely useful for simulation of many industrial MOS and bipolar integrated circuits. In the remainder of this chapter, the relaxation methods are described and their mathematical properties are presented.

3.2.1. Linear Relaxation

Two common linear iterative methods are the Gauss-Jacobi (GJ) and Gauss-Seidel (GS). The methods differ only in the information they use when solving a particular equation as shown in the two algorithms given below. The superscript **k** is the iteration count, and ε is some small error tolerance.

Notice that in the GJ method each $\mathbf{x}_i^k$ is computed using the iteration values $\mathbf{x}_j^{(k-1)}$, $\mathbf{j}$=1, $\cdots$,$\mathbf{n}$, which are the values from the previous iteration. In the GS method, the latest iteration values are used as soon as they become available. The **forall** construct in Algorithm 3.1 suggests that all

n variables can be computed in parallel during each iteration. The **foreach** construct in Algorithm 3.2 requires that the variables be processed in a particular sequence.

<u>**Algorithm 3.1** (Gauss-Jacobi Method to solve **Ax = b**)</u>

> **k**← 0;
> guess $\mathbf{x}^0$;
> **repeat** {
>> **k**← **k**+1;
>> **forall** (**i** ∈ { 1, ··· ,**n** })
>>
>> $$x_i^k = \frac{1}{a_{ii}}\left[bi - \sum_{j=1}^{i-1} a_{ij}x_j^{k-1} - \sum_{j=i+1}^{n} a_{ij}x_j^{k-1} \right] ;$$
>
> } **until** (| $x_i^k - x_i^{k-1}$ | ≤ ε , **i**=1, ··· ,**n**);

□

<u>**Algorithm 3.2** (Gauss-Seidel Method to solve **Ax = b**)</u>

> **k**← 0 ;
> guess $\mathbf{x}^0$;
> **repeat** {
>> **k**← **k**+1;
>> **foreach** (**i** ∈ { 1,..,**n** })
>>
>> $$x_i^k = \frac{1}{a_{ii}}\left[bi - \sum_{j=1}^{i-1} a_{ij}x_j^{k} - \sum_{j=i+1}^{n} a_{ij}x_j^{k-1} \right] ;$$
>
> } **until** (| $x_i^k - x_i^{k-1}$ | ≤ ε , **i**=1, ··· ,**n**);

□

Linear relaxation schemes are usually described using a *splitting* notation that separates **A** into two components:

$$\mathbf{A} = \mathbf{B} - \mathbf{C}$$

where **B** is a nonsingular matrix such that linear systems of the form $\mathbf{Bx} = \mathbf{d}$ are "easy" to solve. Various relaxation schemes can be constructed by choosing different **B** and **C** matrices in the iterative equation:

$$\mathbf{x}^{k+1} = \mathbf{B}^{-1}\mathbf{C}\mathbf{x}^{k} + \mathbf{B}^{-1}\mathbf{b}$$

In particular, if **A** is decomposed into its diagonal, strictly lower-triangular and strictly upper-triangular parts, D, L and U, respectively such that $\mathbf{A} = \mathbf{L} + \mathbf{D} + \mathbf{U}$, then the GS method is obtained by setting

$$\mathbf{B} = (\mathbf{L}+\mathbf{D}) \qquad \mathbf{C} = -\,\mathbf{U} \tag{3.2}$$

and the GJ method is obtained using

$$\mathbf{B} = \mathbf{D} \qquad \mathbf{C} = -\,(\mathbf{L}+\mathbf{U}). \tag{3.3}$$

Since relaxation methods are iterative, the question naturally arises as to whether or not these methods converge to the correct solution and, if so, under what conditions? The requirements for convergence are stated in the following standard theorem [VAR62]:

Theorem 3.1: Suppose $\mathbf{b} \in \mathbb{R}^{n}$ and $\mathbf{A} = \mathbf{B} - \mathbf{C} \in \mathbb{R}^{n\mathrm{x}n}$ is nonsingular. If **B** is nonsingular and the spectral radius of $\mathbf{B}^{-1}\mathbf{C}$, given by $\rho(\mathbf{B}^{-1}\mathbf{C})$, satisfies the condition $\rho(\mathbf{B}^{-1}\mathbf{C})<1$, then the iterates $\mathbf{x}^{(k)}$ defined by $\mathbf{Bx}^{(k+1)} = \mathbf{Cx}^{(k)} + \mathbf{b}$ converge to $\mathbf{x}^{*} = \mathbf{A}^{-1}\mathbf{b}$ for any starting vector $\mathbf{x}^{(0)}$. □

In other words, the magnitude of the largest eigenvalue of the iteration matrix $\mathbf{B}^{-1}\mathbf{C}$ must be strictly less than 1 to guarantee convergence of a linear relaxation method. A condition which guarantees that $\rho(\mathbf{B}^{-1}\mathbf{C}) < 1$ is if **A** is strictly diagonally dominant. A matrix has this property if the diagonal term in each row **i** is greater than the sum of the off-diagonal terms in the same row, i.e.,

$$\sum_{\substack{j=1 \\ j \neq i}}^{n} |a_{ij}| < |a_{ii}| \quad \text{for } 1 \leq i \leq n$$

and the "more dominant" the diagonal, the more rapid will be the convergence. This condition can be satisfied if each node in the circuit has a ground capacitor(C_i). Then, when numerical integration method is applied to discretize the time derivative, as describe in Section 2.3, $a\frac{C_i}{h_n}$ is added to each diagonal term. Here, h_n is the time-step and a is some constant that depends on the numerical integration method used. Thus, by choosing a small enough h_n, Theorem 3.1 can be satisfied.

While the rate of convergence of these linear relaxation methods is linear, a number of techniques are available to improve the convergence speed. For example, in the GS method, the order in which the equations are solved usually has a strong effect on the number of iterations required to converge. Consider the case when matrix A is lower triangular. If processed in the sequence, $x_1, x_2, \cdots, x_n$, then one relaxation iteration is sufficient to obtain the correct solution. However, if processed in the reverse order, then n iterations are required to obtain the correct solution. Therefore, equation ordering is usually performed on the variables whenever GS is used.

Another technique to improve convergence, also used in conjunction with the Gauss-Seidel method, is the method of Successive Overrelaxation (SOR). In this approach, the Gauss-Seidel method is used initially to generate an intermediate value, $\tilde{x}_i^{(k+1)}$, using the equation

$$\tilde{x}_i^{(k+1)} = B^{-1}Cx_i^{(k)} + B^{-1}b$$

where B and C are defined by Eq. (3.2). The actual value of $x_i^{(k+1)}$ is obtained by taking a weighted combination of the previous iteration and the intermediate value which depends on a relaxation parameter, ω.

$$x_i^{(k+1)} = (1-\omega)x_i^{(k)} + \omega\tilde{x}_i^{(k+1)}$$

The SOR method can also be defined in terms of the splitting notation with $\mathbf{B} = \omega^{-1}(\mathbf{D}+\omega\mathbf{L})$, and $\mathbf{C} = \omega^{-1}[(1-\omega)\mathbf{D}-\omega\mathbf{U}]$. While the proper choice of ω can greatly reduce the number of iterations, an optimal value of ω can only be computed *a priori* for a limited number of cases. In general, it may be necessary to perform a somewhat complicated eigenvalue analysis to determine the best value of ω. In practice, adaptive algorithms are used to select an appropriate value for ω during the solution process.

Linear relaxation methods can be used in conjunction with the solution of nonlinear equations to solve the linear systems generated by Newton's method. For example, the Newton-SOR method is a combination of the Newton-Raphson method and the SOR method. In this composite algorithm, the Newton iteration can be considered as the "outer loop" and the SOR iteration as the "inner loop." While it is possible to carry the inner loop to convergence, there is no requirement to do so, as long as the outer loop is iterated to convergence. In general, an **m**-step Newton-SOR method can be defined where **m** is the number of iterations used in the inner loop. For the case **m**=1, a one-step Newton-SOR method is obtained. The Newton-SOR method is only one example of the possible combinations of nonlinear iterative methods and linear iterative methods. For example, Newton's method may be replaced by the secant method and the SOR iteration may be replaced by one of the standard Gauss-Seidel or Gauss-Jacobi methods.

3.2.2. Nonlinear Relaxation

The basic idea of relaxation can also be extended to solve systems of nonlinear equations of the form $\mathbf{F}(\mathbf{x}) = 0$, where $\mathbf{F}:\mathbb{R}^n \rightarrow \mathbb{R}^n$, with components $\mathbf{f}_1, \mathbf{f}_2, \ldots, \mathbf{f}_n$ and $\mathbf{f}_i:\mathbb{R}^n \rightarrow \mathbb{R}$. That is, rather than solving the system using direct matrix techniques, the nonlinear equations can be solved in a decoupled fashion. Two such algorithms are given below. The index **k** is the iteration count, while ε_1 and ε_2 are error tolerances.

Algorithm 3.3 (Nonlinear Gauss-Jacobi Method to solve $\mathbf{F}(\mathbf{x}) = 0$)

> $k \leftarrow 0$; guess $\mathbf{x}^0$;
> **repeat** {
> $k \leftarrow k+1$;
>   **forall** ($i \in \{1, \cdots, n\}$)
>   solve $f_i(x_1^{k-1}, \cdots, x_{i-1}^{k-1}, x_i^k, x_{i+1}^{k-1}, \cdots, x_n^{k-1})=0$ for x_i^k ;
> } **until** ($|x_i^k - x_i^{k-1}| \le \varepsilon_1$, $|f_i(\mathbf{x}^{k,i})| \le \varepsilon_2$, $i=1, \cdots, n$);

□

Algorithm 3.4 (Nonlinear SOR Method to solve $\mathbf{F}(\mathbf{x}) = 0$)

> $k \leftarrow 0$; guess $\mathbf{x}^0$;
> **repeat** {
> $k \leftarrow k+1$;
>   **foreach** ($i \in \{1, \cdots, n\}$)
>   solve $f_i(x_1^k, \cdots, x_{i-1}^k, x_i^k, x_{i+1}^{k-1}, \cdots, x_n^{k-1}) = 0$ for x_i^k ;
> $x_i^k \leftarrow (1-\omega)x_i^{k-1} + \omega(x_i^k)$;
> } **until** ($|x_i^k - x_i^{k-1}| \le \varepsilon_1$, $|f_i(\mathbf{x}^{k,i})| \le \varepsilon_2$, $i=1, \cdots, n$);

□

In the above algorithms, $\mathbf{x}^{k,i} = (x_1^k, \cdots, x_{i-1}^k, x_i^k, x_{i+1}^k, \cdots, x_n^k)$.

These algorithms are referred to as nonlinear relaxation methods. The steps are very similar to linear relaxation as given in Algorithms 3.1 and 3.2 except that, in this case, each equation in the inner loop is nonlinear. To solve each one-dimensional nonlinear problem, $f_i(\mathbf{x}) = 0$, an iterative technique such as the Newton method or secant method must be used since, in general, a closed-form solution cannot be obtained. Combining the SOR method with the Newton method results in the SOR-Newton algorithm. The general case is the **m**-step SOR-Newton method, where **m** is the number of Newton iterations taken in the inner loop. The question again arises as to the number of inner loop iterations to use.

It can be shown that the rate of convergence of the one-step SOR-Newton method is the same as for the one-step Newton-SOR method [ORT70]. The **m**-step SOR-Newton method also has the same rate as the one-step method implying that it is not worthwhile to take more than one Newton step since the convergence rate is not affected. However, the convergence rate of the **m**-step Newton-SOR method is **m** times the rate of convergence of the one-step method. Therefore, based on the rates of convergence, one might be inclined to choose the **m**-step Newton-SOR to solve a system of nonlinear equations. There is, however, a hidden cost if the partial derivatives are expensive to calculate. Each step of SOR-Newton requires the evaluation of each $\mathbf{f_i}$ and $\mathbf{n}$ partial derivatives, $\frac{\partial \mathbf{f_i}}{\partial \mathbf{x_i}}$. whereas the **m**-step Newton-SOR method requires the evaluation of **f** and all partial derivatives. Based on both operation counts and the rates of convergence given above, the one-step SOR-Newton method appears to be the most efficient and for this reason it is used in Iterated Timing Analysis (ITA) [SAL83]. Note that this implies one iteration in the inner loop. The outer loop is iterated until convergence is obtained. SOR-Newton also offers one additional advantage over Newton-SOR in that waveform latency can be exploited easily. This feature is described in more detail in the chapter to follow.

In a general-purpose implementation of these methods, the iterative process must be terminated when the solution is close enough to $\mathbf{x}^*$. Often, this condition is checked using the test $|\mathbf{x_i^k} - \mathbf{x_i^{k-1}}| \leq \varepsilon_1$. However, this check of convergence is not sufficient in the nonlinear case. A second test is necessary to ensure that each function, $\mathbf{f_i}$, is close enough to zero, and this is specified using the test $|\mathbf{f_i}(\mathbf{x^{k,i}})| \leq \varepsilon_2$ for all $\mathbf{i}$.

The algorithms presented above are meaningful only if the nonlinear equations, which are solved at each step in the inner loop, have unique solutions in some specific domain under consideration. Recall that for linear relaxation, the condition that $\mathbf{a_{ii}} \neq 0$, for all $\mathbf{i}=1, \cdots, \mathbf{n}$ ensures that a solution exists, assuming that the diagonal dominance property holds. A similar condition is required in the nonlinear case. To illustrate this point, let the Jacobian be decomposed into its diagonal, strictly lower-triangular and strictly upper-triangular parts as follows:

$$\mathbf{F}'(\mathbf{x}) = \mathbf{D}(\mathbf{x})+\mathbf{L}(\mathbf{x})+\mathbf{U}(\mathbf{x})$$

The iterations in the nonlinear scheme are well-defined if $\mathbf{F}$ is continuously differentiable in an open neighborhood $\mathbf{S}$ of the point $\mathbf{x}^*$, for which $\mathbf{F}(\mathbf{x}^*)=0$, and $\mathbf{D}(\mathbf{x}^*)$ is nonsingular. The requirements for convergence are also analogous to those for the linear case. By splitting the Jacobian matrix using the previous notation

$$\mathbf{F}'(\mathbf{x}) = \mathbf{B}(\mathbf{x}) - \mathbf{C}(\mathbf{x}),$$

the local convergence of the nonlinear relaxation methods described in Algorithms 3.3 and 3.4 can be stated as follows [ORT70]:

Theorem 3.2: Given $\mathbf{F}:\mathbb{R}^n \rightarrow \mathbb{R}^n$, assume that $\mathbf{F}$ is continuously differentiable in an open neighborhood $\mathbf{S}$ of $\mathbf{x}^*$ and $\mathbf{x}^*$ satisfies $\mathbf{F}(\mathbf{x}^*)=0$. If $\mathbf{B}(\mathbf{x}^*)$ is nonsingular and $\rho(\mathbf{B}(\mathbf{x}^*)^{-1}\mathbf{C}(\mathbf{x}^*))<1$, then there exists an open ball $\mathbf{S}^* \subset \mathbf{S}$ such that the nonlinear relaxation methods given in Algorithms 3.3 and 3.4 converge to $\mathbf{x}^*$ for any initial guess $\mathbf{x}^0 \in \mathbf{S}^*$. □

Recall that under the conditions stated in Theorem 3.1, linear relaxation methods converge for any initial guess. However, for the nonlinear case the convergence result is local since the initial guess must be close enough to the final solution to guarantee convergence. The proof of this theorem may be found in the reference [ORT70].

3.2.3. Waveform Relaxation

The relaxation schemes presented above can be also extended to functions spaces to solve systems of differential equations. This class of algorithms is called *Waveform Relaxation* (WR) [LEL82]. The relaxation variables in WR are elments of function spaces, i.e., they are waveforms in the closed interval [0,**T**], whereas for linear and nonlinear relaxation the variables are simply vectors in Euclidean **n**-space. To illustrate the WR algorithm, consider the circuit simulation problem in the form specified in Eq. (2.9). The WR method for solving this system of equations is given in Algorithm 3.5 below.

<u>**Algorithm 3.5** (WR Gauss-Seidel Algorithm for Solving Eq. (2.9))</u>

$\mathbf{k} \leftarrow 0$;
guess waveform $\mathbf{x}^k(\mathbf{t})$; $\mathbf{t} \in [0,\mathbf{T}]$ such that $\mathbf{x}^k(0) = \mathbf{x}_0$;
repeat {
  $\mathbf{k} \leftarrow \mathbf{k}+1$;
  foreach ($\mathbf{i} \in \{1,..,\mathbf{n}\}$) {
    solve

$$\sum_{j=1}^{i} C_{ij}(x_1^k, \cdots, x_i^k, x_{i+1}^{k-1}, \cdots, x_n^{k-1}, u)\dot{x}_j^k +$$

$$\sum_{j=i+1}^{n} C_{ij}(x_1^k, \cdots, x_i^k, x_{i+1}^{k-1}, \cdots, x_n^{k-1}, u)\dot{x}_j^{k-1} +$$

$$f_i(x_1^k, \cdots, x_i^k, x_{i+1}^{k-1}, \cdots, x_n^{k-1}, u) = 0$$

    for ($\mathbf{x}_i^k(\mathbf{t})$; $\mathbf{t} \in [0,\mathbf{T}]$), with the initial condition $\mathbf{x}_i^k(0) = \mathbf{x}_{i_0}$;

  }
} **until** ($\max_{1 \le i \le n} \max_{t \in [0,T]} | \mathbf{x}_i^k(\mathbf{t}) - \mathbf{x}_i^{k-1}(\mathbf{t}) | \le \varepsilon$)

□

Algorithm 3.5 converts the problem of solving a coupled system of **n** first-order ODEs to the problem of solving **n** separate differential equations, each containing a single variable. The outer loop in the algorithm is the Gauss-Seidel iteration which requires that the latest values of the relaxation variables be used to solve each equation in the inner loop. Each equation in the inner loop is a single nonlinear differential equation, and this equation can be solved using any standard numerical integration method.

The convergence of the Waveform Relaxation method is guaranteed under conditions which are similar to the linear and nonlinear cases, as stated in the following theorem [WHI85C]:

Theorem 3.3: If $\mathbf{C}(\mathbf{x}(\mathbf{t}),\mathbf{u}(\mathbf{t})) \in \mathbb{R}^{n \times n}$ of Eq. (2.9) is strictly diagonally dominant uniformly over all $\mathbf{x}(\mathbf{t}) \in \mathbb{R}^n$ and $\mathbf{u}(\mathbf{t}) \in \mathbb{R}^r$ and Lipschitz continuous with respect to $\mathbf{x}(\mathbf{t})$ for all $\mathbf{u}(\mathbf{t})$, then the sequence of waveforms $\{\mathbf{x}^k\}$ generated by the Gauss-Seidel or Gauss-Jacobi WR algorithm will converge uniformly to the solution of Eq. (2.9) in any bounded interval $[0,\mathbf{T}]$, for any initial guess $\mathbf{x}^0(\mathbf{t})$. □

While this theorem guarantees convergence of the WR algorithm, it does not imply anything about the speed of convergence. Although the method usually converges in a few iterations, it has been observed that in test cases with tight feedback loops, the number of iterations required to converge is proportional to the simulation interval [WHI83]. To improve convergence, the simulation interval $[0,\mathbf{T}]$ is usually divided into smaller intervals, $[0,\mathbf{T}_1]$, $[\mathbf{T}_1,\mathbf{T}_2]$, ... , $[\mathbf{T}_{n-1},\mathbf{T}_n]$, called *windows*. Initially, the WR algorithm is applied only in the first window, $[0,\mathbf{T}_1]$, until the waveforms converge. Then a second window, $[\mathbf{T}_1,\mathbf{T}_2]$, is selected and WR is applied within this interval until the waveforms converge. This continues until the entire simulation interval is covered. Note that the WR method converges more rapidly as the window size is made smaller. One advantage of WR is that the time-steps for each of the variables can be chosen independently of one another, but this advantage is compromised if the windows are too small. Therefore, the window size is an important factor which determines the performance of programs which use the WR method.

3.2.4. Partitioning for Relaxation Methods

Relaxation methods are most effective when applied to a system of equations which are "loosely-coupled," that is, where the variables do not depend too strongly on one another. For this type of system, relaxation methods usually converge quite rapidly. The speed of convergence in the linear case is controlled by the spectral radius of the iteration matrix given by $\rho(\mathbf{B}^{-1}\mathbf{C})$ (using the notation of Theorem 3.1); this is usually close to zero for loosely-coupled systems. However, for an arbitrary problem, there is no guarantee that the spectral radius will be small. In fact, in "tightly-coupled" systems, the spectral radius may be very close to 1 which implies slow convergence. This degrades the performance of the relaxation-based methods compared to those for the direct methods.

The precise meaning of loosely-coupled and tightly-coupled can be described using a simple 2x2 matrix problem:

$$\begin{bmatrix} \mathbf{a}_{11} & \mathbf{a}_{12} \\ \mathbf{a}_{21} & \mathbf{a}_{22} \end{bmatrix} \begin{bmatrix} \mathbf{x}_1 \\ \mathbf{x}_2 \end{bmatrix} = \begin{bmatrix} \mathbf{b}_1 \\ \mathbf{b}_2 \end{bmatrix}$$

Assume that the equations have been ordered such that $\mathbf{x}_1$ is solved before $\mathbf{x}_2$. Then, $\mathbf{a}_{21}$ can be considered as a feed-forward term and $\mathbf{a}_{12}$ can be considered as a feedback term. The spectral radius of the iteration matrix for the GS method (see Theorem 3.1) is given by

$$\rho(\mathbf{B}^{-1}\mathbf{C}) = \frac{|\mathbf{a}_{12}\mathbf{a}_{21}|}{|\mathbf{a}_{11}\mathbf{a}_{22}|}$$

and to guarantee convergence, this value must be strictly less than 1. If both $\mathbf{a}_{12}$ and $\mathbf{a}_{21}$ are non-zero, the variables $\mathbf{x}_1$ and $\mathbf{x}_2$ are considered to be coupled. If both $\mathbf{a}_{12}$ and $\mathbf{a}_{21}$ are large, relative to $\mathbf{a}_{11}$ and $\mathbf{a}_{22}$, then $\mathbf{x}_1$ and $\mathbf{x}_2$ are called *tightly-coupled* variables. If both $\mathbf{a}_{12}$ and $\mathbf{a}_{21}$ are small, then $\mathbf{x}_1$ and $\mathbf{x}_2$ are called *loosely-coupled* variables. Note that if either $\mathbf{a}_{21}$ or $\mathbf{a}_{12}$ is zero, then equation ordering has a significant impact on the number of iterations. In fact, if $\mathbf{a}_{21}=0$, then $\mathbf{x}_2$ should be solved before $\mathbf{x}_1$ so that the solution can be obtained in one iteration. A similar argument applies if $\mathbf{a}_{21}$ is very small compared to $\mathbf{a}_{12}$. Therefore, the main objective in reordering is to make the **A** matrix as lower triangular as possible.

When solving large systems, the definitions given above can be used to partition the system into groups of tightly-coupled variables. Rather than using relaxation methods to solve the tightly-coupled variables within each "block," it is better to solve them using direct methods. The relaxation method can be applied between the blocks, which are loosely-coupled relative to the variables within a block. This gives rise to block relaxation methods [VAR62], which can be viewed as a combination of the direct methods and relaxation methods. As an example, consider the 3x3 matrix problem:

$$\begin{bmatrix} \mathbf{a}_{11} & \mathbf{a}_{12} & 0 \\ 0 & \mathbf{a}_{22} & \mathbf{a}_{23} \\ 0 & \mathbf{a}_{32} & \mathbf{a}_{33} \end{bmatrix} \begin{bmatrix} \mathbf{x}_1 \\ \mathbf{x}_2 \\ \mathbf{x}_3 \end{bmatrix} = \begin{bmatrix} \mathbf{b}_1 \\ \mathbf{b}_2 \\ \mathbf{b}_3 \end{bmatrix}$$

If $\mathbf{x}_2$ and $\mathbf{x}_3$ are tightly-coupled, then many relaxation iterations may be required to solve this problem. However, by grouping $\mathbf{x}_2$ and $\mathbf{x}_3$ into the same block and reordering the variables for the Gauss-Seidel method, the following equation is obtained:

$$\begin{bmatrix} \mathbf{a}_{22} & \mathbf{a}_{23} & 0 \\ \mathbf{a}_{32} & \mathbf{a}_{33} & 0 \\ \mathbf{a}_{12} & 0 & \mathbf{a}_{11} \end{bmatrix} \begin{bmatrix} \mathbf{x}_2 \\ \mathbf{x}_3 \\ \mathbf{x}_1 \end{bmatrix} = \begin{bmatrix} \mathbf{b}_2 \\ \mathbf{b}_3 \\ \mathbf{b}_1 \end{bmatrix}$$

If $\mathbf{x}_2$ and $\mathbf{x}_3$ are solved using direct methods, then this problem can be solved using a single relaxation iteration.

The examples above are only meant to show that proper ordering and partitioning are extremely important in the relaxation-based methods. In Chapter 4, the details of the ordering and partitioning algorithms will be described, including a discussion of the convergence requirements, the handling of MNA elements and other methods to improve simulation speed in the presence of tight coupling.

CHAPTER 4

ITERATED TIMING ANALYSIS

In the previous two chapters, the circuit simulation problem was identified and efficient techniques to solve the problem were described. In this chapter, a detailed description of event-driven electrical simulation based on nonlinear relaxation methods is provided. The chapter begins with the equation flow for nonlinear relaxation when applied to the circuit simulation problem. Then the timing analysis and iterated timing analysis (ITA) algorithms are described. Algorithms for event-driven electrical simulation with block partitioning and global variable time-step control are presented in the next two sections. Finally, the issues relating to latency detection and event scheduling in ITA are discussed.

4.1. EQUATION FLOW FOR NONLINEAR RELAXATION

The starting point for the description is the system of nonlinear differential equations describing the circuit behavior using the charge-based formulation:

$$\dot{\mathbf{q}}(\mathbf{v}(\mathbf{t})) = \mathbf{f}(\mathbf{v}(\mathbf{t}),\mathbf{u}(\mathbf{t})), \quad \mathbf{v}(0)=\mathbf{V}, \quad \mathbf{t} \in [0,\mathbf{T}] \tag{4.1}$$

where $\mathbf{q}$ is the charge associated with the capacitors connected to each node, $\mathbf{f}$ is the sum of the currents charging the capacitances at each node, $\mathbf{u}$ is the set of input voltages and $\mathbf{v}$ is the set of unknown node voltages. Using trapezoidal integration [CHU75] to discretize the system in Eq. (4.1), the following system of nonlinear difference equations is obtained:

$$\mathbf{q}_{n+1} = \mathbf{q}_n + \frac{\mathbf{h}_n}{2}(\mathbf{f}_{n+1} + \mathbf{f}_n) \tag{4.2}$$

where the subscripts $\mathbf{n}$ and $\mathbf{n}$+1 refer to time points $\mathbf{t}_n$ and $\mathbf{t}_{n+1} = \mathbf{t}_n + \mathbf{h}_n$, respectively, and $\mathbf{h}_n$ is the integration step size. This equation can be formulated as a nonlinear problem, as follows:

$$\mathbf{F}(\mathbf{v}) = \frac{2}{\mathbf{h}_n}(\mathbf{q}_{n+1} - \mathbf{q}_n) - (\mathbf{f}_{n+1} + \mathbf{f}_n) = 0 \tag{4.3}$$

Instead of solving this system of equations using standard techniques [NAG75], the strategy in this section is to use nonlinear relaxation. That is, use the Newton method to solve each equation in the system separately and a relaxation method to guarantee that the solutions are mutually consistent. The expression for the **i**th equation in Eq. (4.3) solved using the Newton method is

$$\mathbf{J}_{\mathbf{F}_i}(\mathbf{v}^k)(\mathbf{v}_i^{k+1} - \mathbf{v}_i^k) = -\mathbf{F}_i(\mathbf{v}^k) \tag{4.4}$$

where the index **k** is the iteration counter for the Newton method and $\mathbf{J}_{\mathbf{F}_i}(\mathbf{v})$ is the **i**th diagonal term of the Jacobian matrix of $\mathbf{F}(\mathbf{v})$ given by

$$\mathbf{J}_{\mathbf{F}_i} = \frac{2}{\mathbf{h}}\frac{\partial \mathbf{q}_i(\mathbf{v}^k)}{\partial \mathbf{v}_i} - \frac{\partial \mathbf{f}_i(\mathbf{v}^k)}{\partial \mathbf{v}_i} \tag{4.5}$$

Usually a number of iterations are required to obtain the correct solution. However, in this case, since a converged relaxation method is used to guarantee a consistent solution to the system of equations, the Newton iteration for each equation need not be carried to convergence. In fact, from an efficiency standpoint, *only one iteration* should be used to approximate the solution of each equation before moving to the next equation, as described earlier in Chapter 3. The resulting one-step Gauss-Seidel-Newton relaxation algorithm is specified precisely in the following, using the definition:

$$\mathbf{v}^{k,i} = [\mathbf{v}_1^{k+1}, \mathbf{v}_2^{k+1}, \cdots, \mathbf{v}_{i-1}^{k+1}, \mathbf{v}_i^k, \mathbf{v}_{i+1}^k, \cdots, \mathbf{v}_n^k]^T$$

where the superscript **T** denotes the transpose of a vector. This definition is based on the Gauss-Seidel method which uses the $\mathbf{k}+1^{st}$ values of all other components, whenever possible, in computing the $\mathbf{k}+1^{st}$ value of $\mathbf{v}_i$.

Here **n** is the number of equations in the system. The algorithm for a one-step Gauss-Seidel-Newton Relaxation method is given below:

Algorithm 4.1 (Gauss-Seidel-Newton Relaxation Method)

repeat {

  foreach ($\mathbf{i} \in \{1, \cdots, \mathbf{n}\}$) {

    solve $\mathbf{J}_{\mathbf{F}_i}(\mathbf{v}^{k,i})(\mathbf{v}_i^{k+1} - \mathbf{v}_i^k) = -\mathbf{F}_i(\mathbf{v}^{k,i})$ for $\mathbf{v}_i^{k+1}$

    where $\mathbf{F}_i(\mathbf{v})$ is specified in Eq. (4.3) and

      $\mathbf{J}_{\mathbf{F}_i}(\mathbf{v})$ is specified in Eq. (4.5) ;

  }

} **until** ($\|\mathbf{v}_i^{k+1} - \mathbf{v}_i^k\| < \varepsilon_1$, $\|\mathbf{F}_i\| < \varepsilon_2$, $\mathbf{i}=1,..,\mathbf{n}$)

□

4.2. TIMING ANALYSIS ALGORITHMS

The first published program to use techniques based on nonlinear relaxation for circuit simulation was the MOTIS program [CHA75]. It used backward-Euler integration, a Gauss-Jacobi-Newton relaxation algorithm, and node-by-node decomposition (that is, it solved for one node voltage at a time). In MOTIS, a simple modification was made to the relaxation scheme based on the conjecture that there exists a small enough time-step, $\mathbf{h}_{min}$, such that the method obtains the correct solution in exactly one iteration. At each time point, $\mathbf{t}_{n+1}$, the program computed new values of all node voltages using only one iteration of the Gauss-Jacobi-Newton method and accepted the results as the correct solutions at $\mathbf{t}_{n+1}$. It was believed that iterating the outer relaxation loop to convergence would be both expensive and unnecessary for most MOS logic circuits. However, the resulting accuracy of this approach relied heavily on three things:

1) The user's ability to select an appropriate time-step based on knowledge of the circuit characteristics.

2) The fact that the global error reduces to zero when a node voltage reaches the supply voltage or ground.

3) Only a limited number of well-characterized circuit topologies (CMOS polycells) were used to build a design.

The initial speed improvements obtained using this approach were extremely encouraging, partially due to the simplified numerical techniques and partially due to the use of table lookup models for the MOS devices. The combined techniques were shown to be over two orders of magnitude faster than standard techniques when applied to large digital MOS circuits [CHA75]. Since the method was intended to provide first-order timing information of MOS logic circuits, it was called "Timing Analysis" or "Timing Simulation."

Although timing analysis provided an electrical simulation capability with execution speeds comparable to logic simulation, it had a number of problems. For example, the choice of a proper time-step to guarantee accurate solutions was very difficult to determine in general. In addition, the method had severe accuracy problems for circuits containing elements such as large floating capacitors[1], small floating resistors and transfer gates. The MOTIS program avoided this problem for floating capacitors by not allowing them in the circuit description and solved collections of transfer gates using direct methods.

A number of improvements to the basic technique was suggested to overcome the inherent accuracy limitations of the method. In particular, the MOTIS-C program [FAN77] employed trapezoidal integration and one iteration of the Gauss-Seidel-Newton relaxation algorithm. Since timing analysis algorithms based on the Gauss-Seidel principle use updated information at $\mathbf{t}_{n+1}$ whenever possible, the accuracy is generally better than one

[1] A "floating" element is a two-terminal device whose terminals are not connected to either ground or to a power supply.

based on the Gauss-Jacobi method. The simulation time-step was selected automatically in the program by doing a simple analysis of the time constants associated with each node and by using some fraction of the smallest time constant as the step size. However, MOTIS-C still suffered from problems similar to those for MOTIS.

A modified timing analysis algorithm was implemented in SPLICE1.3 [NEW78] as part of a mixed-mode simulation capability. Although backward-Euler integration was used in this program, a number of other noteworthy enhancements were made to the underlying timing analysis algorithm. The first enhancement was based on two observations:

1) Most of the node voltages in a large digital circuit remain stationary at a given time point (the latency property). Computing the solution for these nodes is unnecessary.

2) The order in which the nodes are solved has a strong influence on the accuracy of the solution for timing analysis algorithms based on the Gauss-Seidel principle.

These observations suggested that a good strategy would be to identify the "active" nodes at each time point and process these nodes in an order based on the direction of signal flow. In SPLICE1.3, a single mechanism was used to perform both tasks: an *event-driven, selective-trace* algorithm normally associated with logic simulation [SZY75]. This mechanism is described in the following paragraphs.

The SPLICE1 program treats a circuit as a signal-flow graph and constructs a corresponding directed graph for the circuit given by **G=G(X,E)**, where **X** is the set of vertices and **E** is the set of directed edges of the graph. Two tables, the *fanin* and *fanout* tables, are constructed at each vertex based on the following definitions:

Definition 4.1 (Fanin and Fanout nodes)

A node $\mathbf{x_k}$ is called a fanin node of $\mathbf{x_i}$, and is specified as $\mathbf{x_k} \in \mathbf{Fanin}(\mathbf{x_i})$, if $\mathbf{x_k}$ directly affects $\mathbf{x_i}$. A node $\mathbf{x_j}$ is called a fanout node of $\mathbf{x_i}$, and is specified as $\mathbf{x_j} \in \mathbf{Fanout}(\mathbf{x_i})$, if $\mathbf{x_j}$ is directly affected by $\mathbf{x_i}$. □

Whenever the value of an input node or any internal node changes, it is possible to *schedule* all of its fanouts to be processed. In this way the effect of a change at the input to a circuit may be *traced* as it propagates to other circuit nodes via the fanout tables. Since the only nodes that are processed are those which are affected directly by the change, this technique is *selective* and hence its name: *selective trace.* If such a selective trace algorithm is used with the fanout tables, the order in which the nodes are updated becomes a function of the signals flowing in the network and is therefore a dynamic ordering.

To make the processing efficient, and for consistency with the logic simulator in the SPLICE1 program, the total simulation period, $\mathbf{T_{stop}}$, is divided into uniform steps, referred to as the *Minimum Resolvable Time* (**mrt**). A time queue is constructed and the time slots in this queue define distinct points in time separated by one **mrt**. Hence, events are scheduled at integer multiples of **mrt** in the queue. The simple event scheduling approach used in SPLICE1 for timing analysis is given in Algorithm 4.2. The routine *NextEventTime*(**t**) examines successive time slots in the time queue starting at time **t** and returns the next time point where one or more events have been scheduled. The external input nodes to a circuit are denoted as $\mathbf{e_k}$.

As seen in the algorithm below, three separate event scheduling mechanisms exist:

1) External inputs generate events whenever they make transitions from one value to another,

2) Internal nodes can schedule themselves to be processed, and

3) Internal nodes can schedule their fanout nodes to be processed.

Note that if $\mathbf{x_i}$ is not active, then neither $\mathbf{x_i}$ nor its fanouts are scheduled. However, since nodes may schedule themselves, the fanouts of $\mathbf{x_i}$ may still be active even though $\mathbf{x_i}$ is not. The importance of this fact and other issues associated with electrical event scheduling will be presented in Section 4.5. Also, the precise meaning of "active" is elaborated further in Section 4.5.

<u>**Algorithm 4.2** (Event Scheduling Algorithm in SPLICE1)</u>

```
tn← 0;
while ( tn≤Tstop) {
    tn← NextEventTime( tn );
    foreach ( input k at tn ) {
        if ( ek is "active" )
            forall ( xj ∈ Fanout(ek) ) schedule( xj, tn );
    }
    foreach ( event i at tn ) {
        process node xi by computing xi(tn);
        if ( xi is "active" ) {
            schedule( xi, tn+h );
            forall ( xj ∈ Fanout(xi) ) schedule( xj, tn );
        }
    }
}
```
□

The use of event-driven, selective-trace techniques give greatly improved accuracy of SPLICE1.3 compared to those for the MOTIS and MOTIS-C programs. In addition, a further improvement was realized using a variable time-step control, as follows. Initially, every node is solved using a common step size given by the **mrt**. If the change in either the voltage at a node or the current through any device connected to the node is large, its solution is recomputed in the **mrt** interval using smaller steps and a single iteration at each time point. Each of the smaller steps may be further refined to insure that the changes in voltage and current are within acceptable limits. Therefore, the local time-steps for each node are based on limiting change of the node voltage and its associated currents over each step[2]. While the run time was noticeably higher, this variable time-step control was extremely effective in improving the accuracy of the results.

Other enhancements were developed in SPLICE1.3 to handle tightly-coupled circuits. SPLICE1.3 used the Implicit-Implicit-Explicit (IIE) method [NEW80] to handle floating capacitors. To accommodate large blocks of tightly-coupled circuit elements, the program allowed the user to define "circuit" blocks. These blocks would be solved using standard direct matrix techniques. However, instead of using a single iteration, the Newton iteration in the inner loop was carried to convergence since the elements inside the circuit block were considered to be "highly" nonlinear. However, the outer relaxation iteration was only performed once.

While the results from programs using timing analysis were within acceptable accuracy limits for a certain class of problems, a rigorous mathematical analysis indicated that these methods have inherent stability and accuracy problems [DEM81]. This severely limited the application of the technique. Another problem, cited earlier, was that timing analysis programs relied on the user's knowledge of the underlying algorithms and improper usage could produce the wrong answer. Circuit designers have

[2] Note that a variable time-step control based on local truncation error is not easy to define here since the relaxation loop is not carried to convergence. The local error (i.e., the error over one step) is due to the integration method and the fact that the iteration is not carried to convergence.

been known to lose confidence in a simulator if it occasionally produces the wrong answer, whatever the reason. Therefore, this approach has not been widely accepted, although it is heavily used where the approach has been thoroughly developed, is well-understood, and is applied to a restricted class of circuit topologies.

4.3. SPLICE1.7 - FIXED TIME-STEP ITA

The reluctance to close the outer relaxation loop in timing analysis was primarily due to its perceived high cost. However, the event-driven techniques significantly reduced the cost of timing analysis for large problems since only a small fraction of the nodes is processed at each time point. A number of other improved timing analysis algorithms were proposed [DEM83] but they used at least two iterations or required the use of expensive function evaluations, which increased greatly the cost of the simulation. As described earlier, the variable step approach in SPLICE1.3 improved the accuracy somewhat at the expense of additional iterations. The additional cost was thought to be worthwhile due to the improved reliability.

The next step, naturally, is to close the relaxation loop and examine the true cost of iterating to convergence, given that event-driven selective trace is employed to improve efficiency. This was done in the SPLICE1.6 program, which later evolved to be SPLICE1.7, and the technique was named Iterated Timing Analysis or ITA [SAL83]. The prototype version of ITA used backward-Euler integration, node-by-node decomposition and a fixed time-step based on the **mrt**. The fixed time-step algorithm was kept for consistency with the existing scheduler and logic simulation portions of SPLICE1. The ITA algorithm in SPLICE1.7 is a simple extension of Algorithm 4.2 as shown below.

Algorithm 4.3 (Fixed Time-Step ITA)

```
tn ← 0;
while (tn ≤ Tstop) {
    tn ← NextEventTime( tn );
    foreach ( input k at tn )
        if ( ek is active )
            forall ( vj ∈ Fanout(ei) ) schedule( vj, tn );
    repeat {
        foreach ( event i at tn ) {
            solve J_Fi(v^{k,i})(vi^{k+1} − vi^k) = − Fi(v^{k,i}) for vi^{k+1}
             where Fi(v) is specified in Eq. (4.3) and
                  J_Fi(v) is specified in Eq. (4.5) ;
            if ( |vi^{k+1} − vi^k| < ε1,  |Fi| < ε2 ) { /* converged? */
                if ( vi did not converge on last iteration ) {
                    if ( vi is active ) {
                        /* this is the selective-trace portion */
                        schedule( vi, tn+1 );
                        forall ( vj ∈ Fanout(vi) )
                            schedule( vj, tn );
                    }
                    else {/* do nothing (latency) */}
                }
                else {/* do nothing (break feedback loops) */}
            }
            else { /* node has not converged */
                schedule( vi, tn );
                forall ( vj ∈ Fanout(vi) ) schedule( vj, tn );
            }
        }
    } until ( Q is empty at tn )
}
```

□

The following definition is used in the algorithm:

$$\mathbf{v}^{k,i} = [v_1^{k+1}, v_2^{k+1}, \cdots, v_{i-1}^{k+1}, v_i^k, v_{i+1}^k, \cdots, v_n^k]^T$$

The algorithm above has two features not present in the SPLICE1.3 algorithm:

1) If a node voltage does not converge, the node is rescheduled at the current time point $\mathbf{t_n}$ along with its fanout nodes.

2) All nodes are processed until their voltages converge. When a node converges at $\mathbf{t_n}$, it schedules itself at $\mathbf{t_{n+1}}$ and schedules its fanouts at $\mathbf{t_n}$, if active. However, if it is scheduled again at $\mathbf{t_n}$ by one of its fanins, and converges again, it does not schedule any additional events. This approach breaks feedback loops, since two nodes which are fanouts of each other would schedule each other indefinitely at $\mathbf{t_n}$ if this approach was not used.

The speed improvement obtained by the SPLICE1.7 program compared to that for the SPICE2 program was in the range of 5 to 50 times faster for a number of MOS digital circuits containing up to 1200 transistors [SAL83]. However, the ITA approach required approximately twice as much CPU-time to simulate a circuit compared to SPLICE1.3 which used timing simulation [SAL84]. Again, the improvements in reliability and numerical robustness far outweighed the cost of the increase in run-time.

While the converged relaxation scheme is provably better than the non-iterated approach, it is not without problems. One problem is the speed of convergence. For example, SPLICE1.7 was able to simulate accurately an NMOS operational amplifier but it required more than two times the CPU-time used by SPICE2 [NEW83]. The circuit is a tightly-coupled analog circuit with large forward gain and capacitive feedback and, in this application, the node-by-node decomposition strategy used in SPLICE1.7 is inappropriate. For this same reason, convergence is also very slow in the presence of large floating capacitors and small drain and source resistors, usually found in detailed MOS transistor models. Another problem is due to nonconvergence. Since a fixed time-step is used, the program simply stopped when it was unable to converge to a solution within a specified number of relaxation-Newton iterations. Obviously, a variable step algorithm would resolve this problem and would also allow the solutions to be computed accurately based on a local truncation error criterion. These and other problems were solved in the SPLICE2 and iSPLICE3 programs.

The version of ITA implemented in the iSPLICE3 [SAL89A, ACU90] program differs from SPLICE1.7 in two respects:

1) it uses partitioning to improve the speed of convergence for tightly-coupled circuits.

2) it achieves better accuracy by using an LTE-based time-step control.

The iSPLICE3 program also provides detailed MOS level 1 and MOS level 3 transistor models including a charge-conserving capacitance model.

4.4. CIRCUIT PARTITIONING

The node-based ITA approach used in SPLICE1.7 is not appropriate for circuits with tight coupling between two or more nodes, since the convergence can be very slow in this situation. One reason for this problem is that in computing the new value for a particular node, the relaxation process effectively replaces the fanin nodes with ideal voltage sources of constant value. Therefore, the true Norton equivalent contributions from the fanin nodes are not used in the computation of a new value for the node. SPLICE2 used an improved representation of the neighboring nodes based on a current and conductance model, rather than constant voltage sources, and this approach was called the *coupling method* [KLE84]. This fanin model is approximate since the exact Norton equivalent is expensive to calculate during the simulation as it involves path tracing from each fanin node to all other reachable nodes. While this approach improved the convergence speed on some examples, the technique was heuristic in nature and did not solve the general problem of coupling between more than two nodes in feedback loops.

As was realized in early relaxation-based simulators such as SPLICE1, tightly-coupled subcircuits are better solved using direct methods [NEW78]. However, it is difficult for users to identify tightly-coupled blocks manually, especially when the degree of coupling is a function of time and usually changes frequently over the simulation interval. A more effective approach to the coupling problem is to identify strongly-coupled components in the circuit automatically and to group them together to form subcircuits - a process referred to as *circuit partitioning*. Since the variables associated with the subcircuits are assumed to be tightly-coupled, the subcircuits can each be solved using direct matrix techniques, and the relaxation method can be applied between subcircuits. This technique has been used in conjunction with the Waveform Relaxation algorithm [LEL82, CAR84, WHI85A, MAR85, DUM86, SAV93] with great success. The same approach can be used with nonlinear relaxation to improve convergence as mentioned in Chapter 3. However, there is still a requirement of a capacitor to ground from each node to guarantee convergence, and this limits the type of circuits that can be simulated using relaxation method. In the following sections, an approach is described to extend the application of relaxation methods to virtually any type of circuits.

4.4.1. MNA Elements

Historically, elements such as controlled sources, inductors and floating voltage sources have been difficult to incorporate into nodal analysis based circuit simulators. Modified Nodal Analysis [HO75] evolved as a method of equation formulation that allows these elements to be easily included in direct method simulators such as SPICE. This is accomplished by appending the branch constitutive relations of these so called *MNA elements* to the nodal equations, and by adding extra columns to the circuit matrix so that the needed variables, such as current and charge, are also included. This is the formulation used by almost all general-purpose circuit simulators. However, MNA elements have posed problems for relaxation-based simulators as they typically do not have grounded capacitors on their terminals. Therefore, circuits containing MNA elements do not usually satisfy the sufficiency conditions for convergence described earlier. In fact, they produce +1 and -1 terms in off-diagonal locations in the circuit matrix, which are usually large relative to other terms in the matrix.

For example, a voltage-controlled voltage source (VCVS), with $\mathbf{V}_3 - \mathbf{V}_4 = \mathbf{a}(\mathbf{V}_2 - \mathbf{V}_1)$, has the following matrix template:

$$\begin{array}{c} \\ \mathbf{BE} \\ \mathbf{n}_3 \\ \mathbf{n}_4 \end{array} \begin{array}{c} \mathbf{V}_1\ \mathbf{V}_2\ \mathbf{V}_3\ \mathbf{V}_4\ \ \mathbf{I} \\ \begin{bmatrix} \mathbf{a} & -\mathbf{a} & 1 & -1 & 0 \\ - & - & - & - & -1 \\ - & - & - & - & +1 \end{bmatrix} \end{array}$$

The presence of the large off-diagonal terms, **a**, +1 and −1, may degrade convergence, if convergence occurs at all! However, the theorem to be presented in the next section suggests a way of incorporating MNA elements into relaxation-based simulation programs. This involves devising a partitioning algorithm that will embed an MNA element into a subcircuit, such that all boundary nodes of the subcircuit will have capacitances to ground.

4.4.2. New Sufficient Condition for Convergence

Recently, an improved condition for the convergence of block time-point relaxation methods has been proved stating, in effect, that grounded capacitors are only required at certain nodes in the circuit [DES89]. This theorem establishes new sufficient conditions for the convergence of the Gauss-Seidel-Newton (GSN) relaxation method and is given below:

Theorem 4.1 : Consider a system of partitioned nonlinear equations $\mathbf{F}$ of the form:

$$\mathbf{F}(\mathbf{x}) = 0 \quad \text{and} \quad \mathbf{A}(\mathbf{x}^*) = (\partial \mathbf{F}/\partial \mathbf{x})(\mathbf{x}^*)$$

Let $\mathbf{x}^*$ be the solution of $\mathbf{F}$, and let $\mathbf{F}$ be continuously differentiable at $\mathbf{x}^*$. Assume that the circuit is partitioned into blocks and a partial ordering has been determined so that these blocks are numbered in ascending order. In the Gauss-Seidel approach, a lower-numbered block will be evaluated before a higher-numbered block. Then, a node inside a block is a *feedback node* if its circuit equation is a function of some nodes in a higher numbered block. The corresponding node to which the feedback node is connected in the higher numbered block is a *feedforward node*. Now, assume the following:

1) All principal square submatrices of $\mathbf{A}(\mathbf{x}^*)$ are invertible and their inverses have bounded norms.

2) $\partial \mathbf{J}/\partial \mathbf{x}$ is independent of the time-step and has bounded entries.

3) There is a capacitor from each feedback node to ground.

Then, the iterates in the Gauss-Seidel-Newton relaxation method converge to $\mathbf{x}^*$ if the time-step $\mathbf{h}$ is small enough and if the initial guess is sufficiently close to the solution $\mathbf{x}^*$.

Proof : The proof is presented in [DES89]. □

The basic idea of the theorem is depicted in Fig. 4.1. There are 14 nodes and a number of edges representing connections between the nodes. Assume that the nodes are partitioned into 4 subcircuits, Sub1, Sub2, Sub3 and Sub4, respectively. Then, all resulting feedback nodes (as defined above) must have a capacitance to ground to guarantee convergence. In Fig. 4.1, the feedback nodes are 2, 3, 7 and 10 and they all have a capacitive connection to ground. Now, if the conditions 1) and 2) are satisfied, then the Gauss-Seidel-Newton relaxation method is guaranteed to

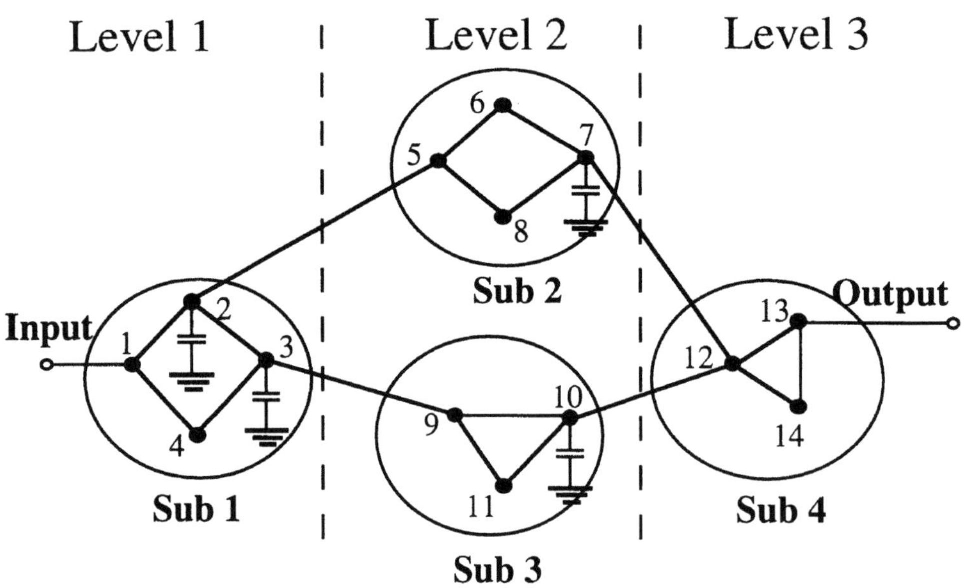

Figure 4.1: Partitioning Approach Suggested by Theorem 4.1

converge if the time-step is small enough. Note that no other capacitors are required in the circuit for convergence. If the Gauss-Jacobi method is used, then the feedforward nodes must also have capacitances to ground. The feedforward nodes in this case are 5, 9 and 12. The event-driven algorithm also requires capacitances to ground at all feedback and feedforward nodes in the circuit, which we refer to as the *peripheral nodes*. The use of this key theorem to expand the scope of relaxation-based simulation is described in the following sections.

4.4.3. A Partitioning Algorithm

A partitioning algorithm is now proposed which improves the convergence of relaxation-based methods and can handle MNA elements. Basically, the partitioner groups tightly-coupled nodes together and ensures that every subcircuit in the network has grounded capacitors at its peripheral nodes. In the case of MNA elements, the partitioning algorithm ensures that all nodes of the device and any internal variables are assigned to the same subcircuit. The pseudo-code for the partitioner is given in Algorithm 4.4.

Algorithm 4.4 (Partitioning MNA Circuits)

```
partition()
{
  nodelist ← {all nodes};
  foreach (MNA element i) {
    place all its nodes in subcircuit i;
    nodelist ← {nodelist} – {MNA nodes};
  }
  G_partitioner(nodelist); /* conductance based */
  C_partitioner(nodelist); /* capacitance based */
  Build_Subcircuit_Data(sublist);          /* tentative subcircuits */
  Levelize(sublist);         /* rank */
```

```
foreach (Level i, i=n, ..., 1) {    /* process in reverse-order */
  foreach (subcircuit j in Level i) {
    levelize(sub_nodelist(j)); /* order nodes in subcircuit */
  }
}
/*Check feedforward and feedback nodes for capacitances*/
foreach (Level i, i=n, ..., 1) {    /*rank*/
  foreach (subcircuit j in Level i) {         /*subcircuit*/
    foreach (node k in subcircuit j) {          /*node*/
      if(k = unvisited) {
        foreach ((x memberof fanout_list(k)) and (x = unmarked))
        {
          Mark(x);
          if(check_feedback(x,k) = NO or
            check_feedforward(x,k) = NO) {
            /*no feedback or feedforward capacitance*/
            Coalesce(j, find_subcircuit(x));
                /*combine subcircuits*/
            Update(j, find_subcircuit(x)); Continue;
          }
        }
      }
      k = visited;
    }
  }
}
Build_subcircuits();
foreach (Level i, i=n, ..., 1) {
  Partial_order(); /* establish final ordering */
}
}
```

□

The first step in the partitioner is to construct a directed graph $G=(V,E)$ where V is the set of nodes in the circuit and E is the set of edges, defined by $E=\{(a,b)\}$. An edge arises whenever the node voltage or node current at node b is dependent on the node voltage or node current at node a. For each such ordered pair, a directed arc from a to b is constructed. An attribute is associated with each node in the graph that indicates whether or not the node has a purely capacitive path to ground. A second attribute is used to indicate whether or not a node has MNA elements attached to it. If so, a unique tag is associated with each such element. This information enables the partitioner to identify all nodes attached to a distinct MNA element. Fig. 4.2 shows an example of a circuit graph to be processed by the partitioner. The graph contains four nodes with the label

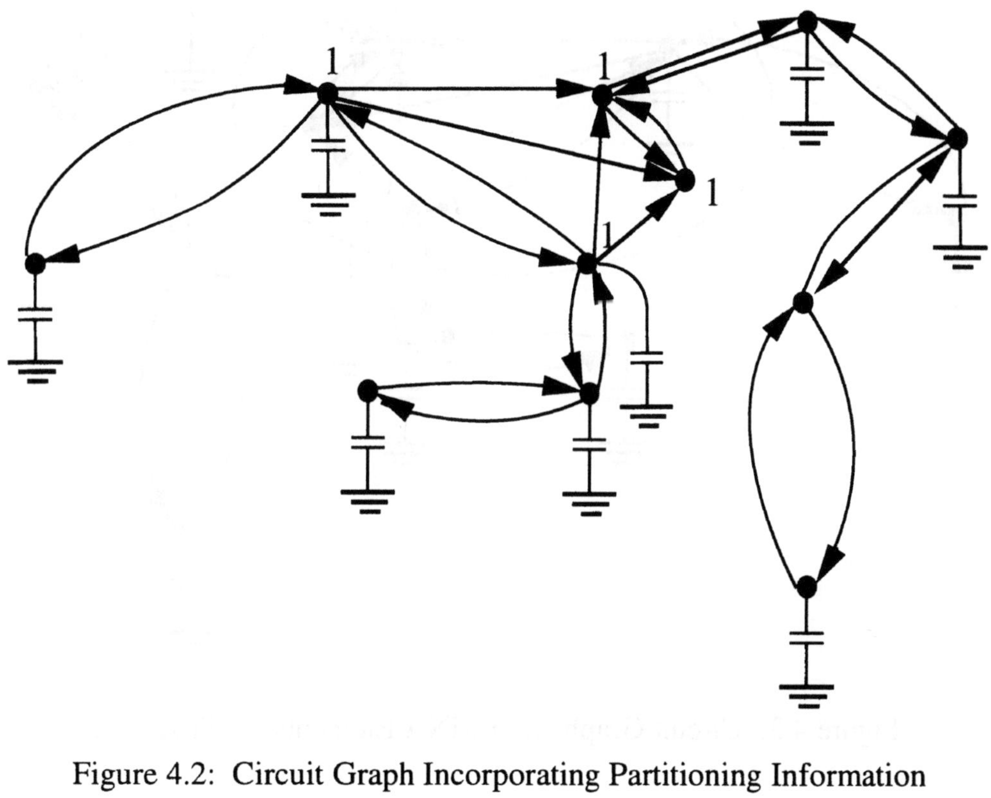

Figure 4.2: Circuit Graph Incorporating Partitioning Information

"1", indicating that they are attached to a common MNA element.

In Algorithm 4.4, all the nodes in the circuit are initially placed in a data structure called a *nodelist*. When the nodes belonging to an MNA element are assigned to a subcircuit, they are removed from the *nodelist*, although they may be revisited by neighboring nodes during the partitioning process. Fig. 4.3 shows the circuit graph of Fig. 4.2 after the MNA element nodes have been assigned to a subcircuit. Next, a static partitioning approach, similar to the one used in the RELAX2 program [WHI85C],

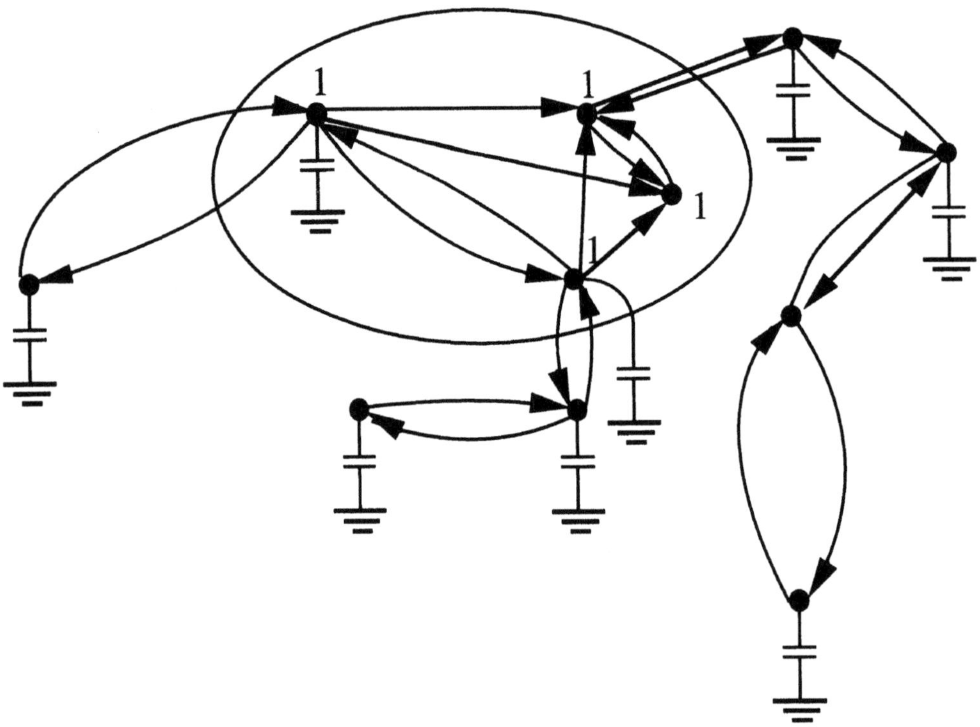

Figure 4.3: Circuit Graph After MNA Elements are Processed

is applied to the rest of the nodes in the *nodelist*. The main goal of static partitioning is to speedup the convergence process of relaxation methods by grouping tightly-coupled nodes together. Recall from Chapter 3 that the speed of convergence is controlled by the contraction factor, γ_∞, in the following way:

$$\|\mathbf{x}^{k+1} - \mathbf{x}^{k}\| \leq \gamma_\infty \|\mathbf{x}^{k} - \mathbf{x}^{k-1}\|$$

For a linear problem, this iteration factor can be computed quite easily. For example, if the linear problem $\mathbf{Ax=b}$ is solved using the Gauss-Seidel algorithm, γ_∞ is equal to the largest eigenvalue of the iteration matrix $[(\mathbf{L+D})^{-1}(-\mathbf{U})]$, where $\mathbf{A=L+D+U}$. Therefore, a two-node linear circuit, such as the one in Fig. 4.4, has an iteration factor (for the conductance portion only) given by:

$$\gamma_\infty = \frac{g_{12}}{(g_2+g_{12})} \frac{g_{12}}{(g_1+g_{12})}$$

A similar expression exists for the capacitance portion of the circuit. Note that if the two nodes are part of a larger circuit, the values of $\mathbf{g_1}$ and $\mathbf{g_2}$ are the Norton equivalent conductances seen from each node looking back into the rest of the circuit.

The static partitioning algorithm makes use of the iteration factor to decide whether or not two nodes should be placed in the same subcircuit. If the factor is close to one and the nodes are solved independently, the convergence would be very slow. Therefore, the two nodes should be placed in the same subcircuit. However, if the factor is close to zero, they may be placed in different subcircuits without adversely affecting the convergence speed. A threshold parameter, α, is used to decide whether or not the nodes should be solved together or separately.

A number of approximations are made in computing the iteration factors when partitioning MOS circuits. As MOS circuits are nonlinear,

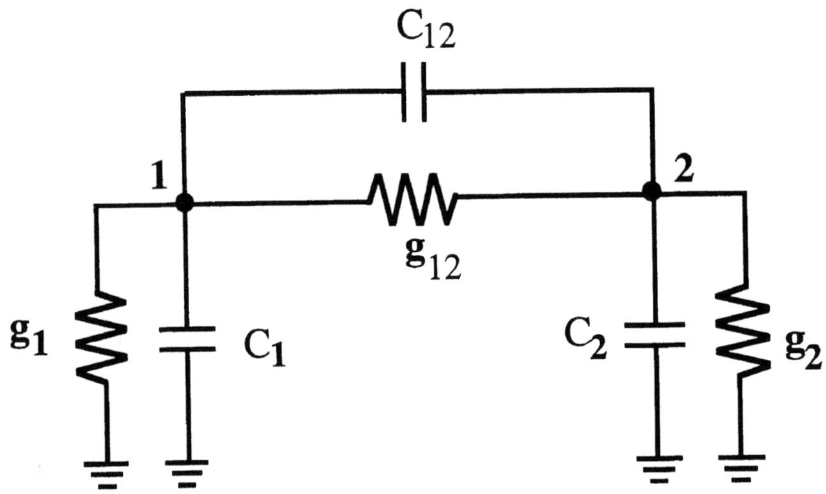

Figure 4.4: Linear Circuit Considered for Partitioning Purposes

each nonlinear device must be replaced by a linear equivalent device. Since a static partitioning strategy is used, worst-case conductance and capacitance values are used when replacing each nonlinear device with a linear one. However, the exact Norton equivalent model seen by each node cannot be computed efficiently because it involves tracing paths from each node to all other nodes in the circuit. For efficiency, the depth of the conductance and capacitance search process is truncated whenever the gate of an MOS transistor is encountered since the conductance of an MOS transistor is zero in the worst case. Path tracing is also terminated when the controlling node of an MNA element (such as a VCVS) having no further connections to other nodes, aside from the controlled nodes, is encountered. Linear inductors and floating-voltage sources are treated as infinite conductance elements during the conductance computing process.

With these heuristics applied, the conductance partitioning approach given in Algorithm 4.5 is obtained. A similar algorithm is used for partitioning based on capacitances.

<u>**Algorithm 4.5** (Conductance Partitioning)</u>

```
G_partitioner(nodelist)
{
   foreach ( node in nodelist ) {
      node 1 ← next node in nodelist;
      foreach ( element connected to node 1) {
         node 2 ← node on other side of element;
         if ( node 1 and node 2 not checked previously ) {
            g12 ←0; g1←0; g2←0;
            foreach ( conductive element between nodes 1 and 2 ) {
                     g12 ←g12 + maximum element conductance over all v;
                     Remove the element from the circuit;
            }
            g1 ← sum of the minimum Norton equivalent
              conductance of each element at node 1
            g2 ← sum of the minimum Norton equivalent
              conductance of each element at node 2
            if ( g12/(g2+g12) g12/(g1+g12) > α ) {
                     Place the two terminal nodes in same subcircuit;
            }
         }
      }
   }
}
```

□

The functions, called *G_partitioner()* and *C_partitioner()* in Algorithm 4.4, carry out the conductance and capacitance partitioning on the nodes in *nodelist* and their immediate neighbors. During this stage, these nodes may be included in subcircuits that contain MNA elements. That is, these nodes may be coupled to nodes that are attached to MNA elements.

After the conductance and capacitance partitioning has been performed, the algorithm builds a preliminary set of subcircuits using the function *Build_Subcircuit_Data()*. Since some of these subcircuits may be coalesced together at later stage, the data-structures for the subcircuits are not finalized at this time. The subcircuits are placed in a *sublist* and are then ranked by the *Levelize()* function.

Fig. 4.5 shows the same circuit graph after the levelizing has been completed. Next, the nodes belonging to a subcircuit are placed in a *sub_nodelist* and ranked. The algorithm then starts with a subcircuit at the highest level and checks each of its peripheral nodes and their associated fanouts, using the functions *check_feedforward()* and *check_feedback()*. If a feedback or feedforward node exists which has no capacitance to ground, the function *Coalesce()* combines the two subcircuits together and assigns the composite subcircuit to the lower level of the two. This procedure ensures that once a node's fanout list has been completely examined, it need never be revisited. In Fig. 4.5, the subcircuit S4 has a feedforward node in subcircuit S5 without a capacitor, so S4 and S5 will be coalesced.

The function *Update()* then assigns a common subcircuit number to all nodes in the composite subcircuit. After the algorithm has stepped through all the peripheral nodes in the circuit, the function *Build_subcircuits()* sets up all the data-structures needed for processing the subcircuits. Lastly, the function *Partial_order()* establishes the order in which subcircuits at a given level must be processed. The ordering ensures that if a feedback connection exists between the subcircuits for a given ordering, the capacitive constraint is satisfied. Fig. 4.6 shows the final subcircuit graph after the partitioning process is complete. Note that

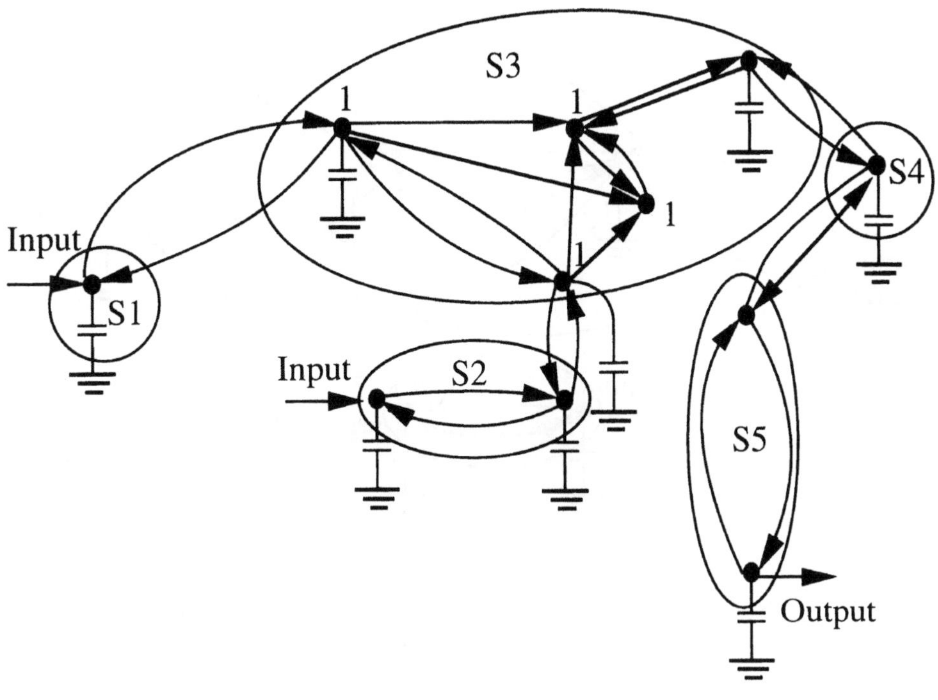

Figure 4.5: Circuit Graph with Levelized Tentative Subcircuits

both S3 and S4 have internal nodes without capacitances to ground, but this circuit will still converge due to the above-mentioned theorem.

Using this partitioning approach, the run times can be reduced significantly compared to the node-based approach. However, the static partitioning strategy described here has a number of problems. The main problem with this approach is that it may produce unnecessarily large subcircuits since worst-case values are used in the partitioning process. The

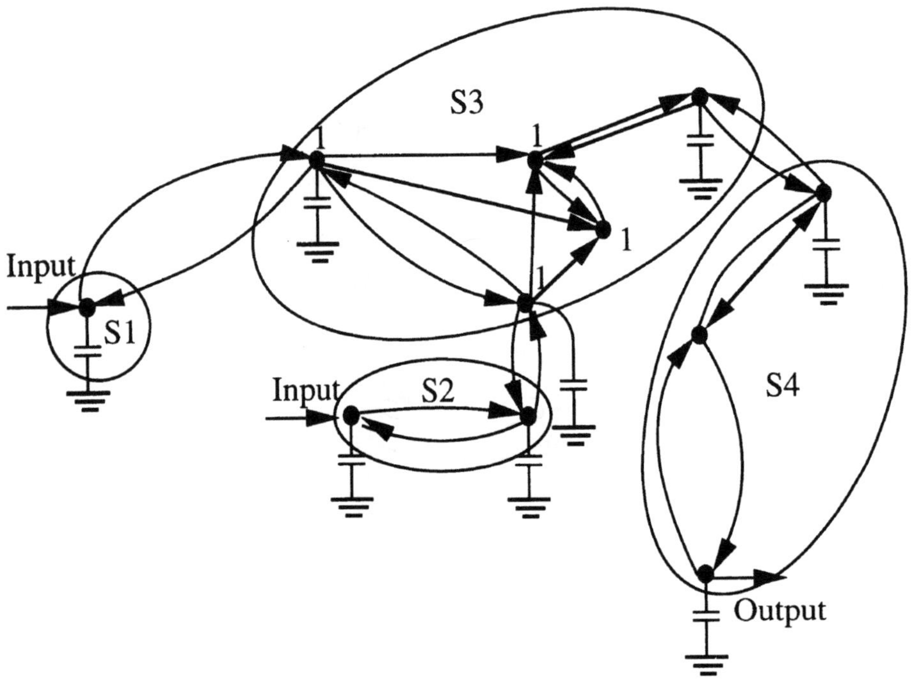

Figure 4.6: Subcircuit Graph After Partitioning is Complete

advantages of the relaxation method are lost if the subcircuits are too large. Latency exploitation is no longer performed at the node level but rather at the subcircuit level. Hence, all nodes in a subcircuit must be latent before the subcircuit is declared latent. While this provides a somewhat stronger condition for latency, it reduces the efficiency of the latency exploitation. Ideally, one would prefer to use small-signal conductance and capacitance values to perform the initial partitioning, and then adjust the subcircuits as these values change during the simulation. This is referred to as *dynamic partitioning* and has already been successfully applied to the simulation of bipolar circuits using Waveform Relaxation [MAR85]. However, this is too expensive to be used in iterated timing analysis.

Another problem is that the partitioning approach given in Algorithm 4.5 is based on a local a criterion that may occasionally make errors. For example, if two nodes are extremely tightly-coupled, relative to their coupling to neighboring nodes, they will be placed in the same subcircuit while the neighboring nodes may be incorrectly placed in different subcircuits. If the neighboring nodes are actually coupled to either of the two external nodes, the convergence will still be slow [WHI85C]. One other problem in partitioning is that it is a time-consuming task. Care must be taken in the definition of the data structures and partitioning algorithms so that the partitioning phase does not dominate the total run time for large circuits. This is more of a concern in dynamic partitioning [MAR85] where the partitioning operation may be performed frequently during the simulation.

4.5. GLOBAL VARIABLE TIME-STEP CONTROL

iSPLICE3 uses a global-variable time-step algorithm in which the components in the system are integrated using a single common time-step. This integration time-step is selected based on the fastest changing variable in the system, the same strategy used in direct methods. However, only the active subcircuits are processed at each time point, and these subcircuits are identified using the selective-trace algorithm. The main steps in the global time-step ITA algorithm are given below following a brief description of the notation to be used.

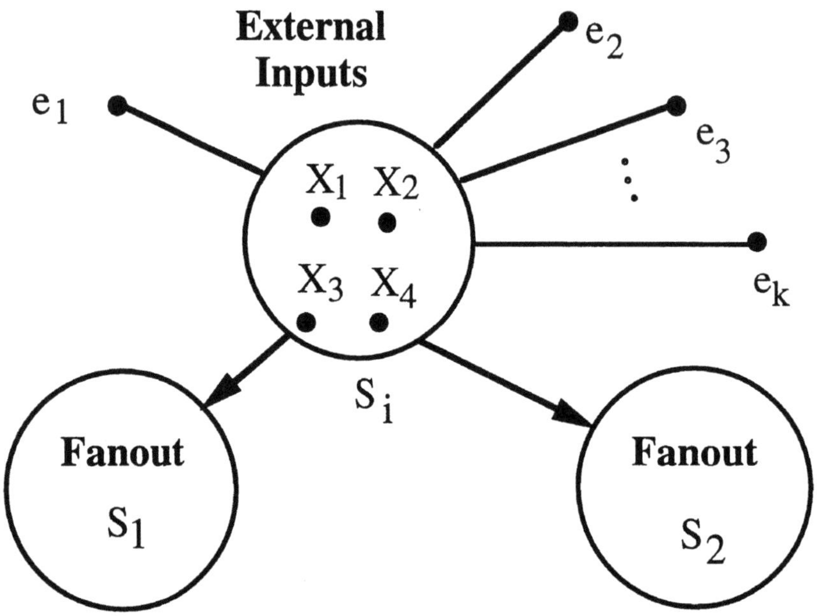

Figure 4.7: Notation Associated with Subcircuits

Notation for Algorithm 4.6 (see Fig. 4.7)

Assume that a given circuit is partitioned into **n** subcircuits $\mathbf{S_1, S_2, \cdots, S_i, \cdots, S_n}$. The **i**th subcircuit, $\mathbf{S_i}$, has $\mathbf{n_i}$ internal variables and $\mathbf{n_e}$ external inputs. The internal variables given by int($\mathbf{S_i}$) = { $\mathbf{x_1}$, $\mathbf{x_2}$, ... , $\mathbf{x_{n_i}}$ } are those variables computed whenever subcircuit $\mathbf{S_i}$ is processed. They are defined in vector form as $\mathbf{v_i=[x_1,x_2,\cdots,x_{n_i}]^T}$. The external inputs of a subcircuit are other nodes which affect the internal nodes of the subcircuit. They are specified as **Fanin**($\mathbf{S_i}$) = { $\mathbf{e_1,e_2,\cdots,e_{n_e}}$}. The fanouts of a subcircuit are associated with the internal nodes of the subcircuits. Hence, the set of subcircuits affected by an internal node, $\mathbf{x_j}$, are specified as **Fanout**($\mathbf{x_j}$) = { $\mathbf{S_1,S_2,\cdots,S_k}$}. The following definition is also used:

$$\mathbf{v}^{k,i} = [\mathbf{v}_1^{k+1}, \mathbf{v}_2^{k+1}, \cdots, \mathbf{v}_{i-1}^{k+1}, \mathbf{v}_i^{k}, \mathbf{v}_{i+1}^{k}, \cdots, \mathbf{v}_n^{k}]^T.$$

<u>**Algorithm 4.6** (Global-Variable-Time-Step ITA)</u>

```
partition();
tn← 0; hmin← hstart;
while ( t ≤ Tstop) {
    stepRejection = FALSE;
    hnext← hmin; tn← tn + hnext; hmin← hmax;
    foreach ( input ik at tn )
        if ( ek is active )
            forall ( Sj ∈ Fanout(ek))  schedule( Sj, tn );
    repeat {
        foreach ( event i at tn ) {
            solve JFi(v^k,i)(vi^k+1 – vi^k) = – Fi(v^k,i) for vi^k+1
              corresponding to subcircuit Si;
            if ( ||vi^k+1 – vi^k||<ε1, ||Fi||<ε2 ) { /*converged? */
              if ( vi did not converge on last iteration ) {
                foreach ( xi ∈ int(Si) ){
                  if ( xi is active ) {
                    if ( CheckAccuracy( xi ) = TRUE ) {
                      hi← pickStep( xi );
                      hmin← min( hmin, hi );
                      schedule( xi, tn+1 );
                      forall (Sj ∈ Fanout(xi))
                        schedule( Sj, tn );
                    }
                    else { /* reject solution */
                        tn← tn– hmin;
                        hmin← hmin/2;
                        stepRejected = TRUE;
                    }
                  }
```

```
                }
            }
        }
        else {  /* subcircuit has not converged yet */
            if ( itercnt > maxitercnt ) {
                tn← tn− hmin; hmin← hmin/2;
                stepRejected = TRUE;
            }
            else {
                schedule( Si, tn );
                foreach ( xi ∈ int(Si) ) {
                    if ( xi is active )
                        forall ( Sj ∈ Fanout(xi) )
                            schedule( Sj, tn );
                }
            }
        }
    }
    } until (( Q is empty at tn ) OR (stepRejection) )
}
```

□

In the algorithm above, the *CheckAccuracy(***x***)* routine uses a local truncation error criterion to determine if the computed solution for **x** is accurate and, if so, returns "TRUE." The *PickStep(***x***)* routine uses an LTE estimate to pick the next recommended step size for **x**.

The main differences between this algorithm and the one used in SPLICE1.7 are due to the actions taken when the subcircuit variables converge at a time point and when they do not converge in a specified number of relaxation-Newton iterations. When the active subcircuits converge at a time point, $\mathbf{t_n}$, the local truncation errors for their internal variables are estimated [BRA72] and the new global time-step, $\mathbf{h_{next}}$, is set to the smallest recommended step in the system, $\mathbf{h_{min}}$. If the accuracy in the solution computed at $\mathbf{t_n}$ is unacceptable, the solution is rejected and the integration is retried with the smaller time-step. Similarly, if the iterations do not converge within a specified number of iterations, the time-step is rejected and a smaller step is used.

4.6. ELECTRICAL EVENTS AND EVENT SCHEDULING

4.6.1. Latency Detection

The most critical aspect in ITA, in terms of accuracy, is the detection of the latency condition. For example, if component **x** is identified as being latent prematurely, any small errors in its value will be propagated to the other components producing errors in their solutions. If the component is thought to be latent but, in reality, it is changing very slowly, the results may be completely wrong. Then the overriding question is: how can one be sure that a variable has reached a steady-state value? The simplest approach is to test if the following condition is satisfied:

Latency Condition 1:

$$|\mathbf{x}_{n+1} - \mathbf{x}_n| < \varepsilon_x \tag{4.6}$$

where $\mathbf{x}_{n+1} = \mathbf{x}(\mathbf{t}_{n+1})$, $\mathbf{x}_n = \mathbf{x}(\mathbf{t}_n)$ and ε_x is some small number. As illustrated in Fig. 4.8, the component is considered latent if the difference in the computed solution at two successive time points is less than some prespecified amount, ε_x. For a fixed time-step ITA algorithm [SAL83], this is a reasonable check as long as ε_x is specified properly and one additional check is done, as described shortly. There are situations where Condition 1 may fail, as shown in Fig. 4.9, where the true solution rises and then falls before reaching a steady-state value. If the time points are chosen such that Condition 1 is satisfied, latency will be detected incorrectly. A more conservative version of Condition 1 requires that the inequality be satisfied for two time points that are not adjacent.

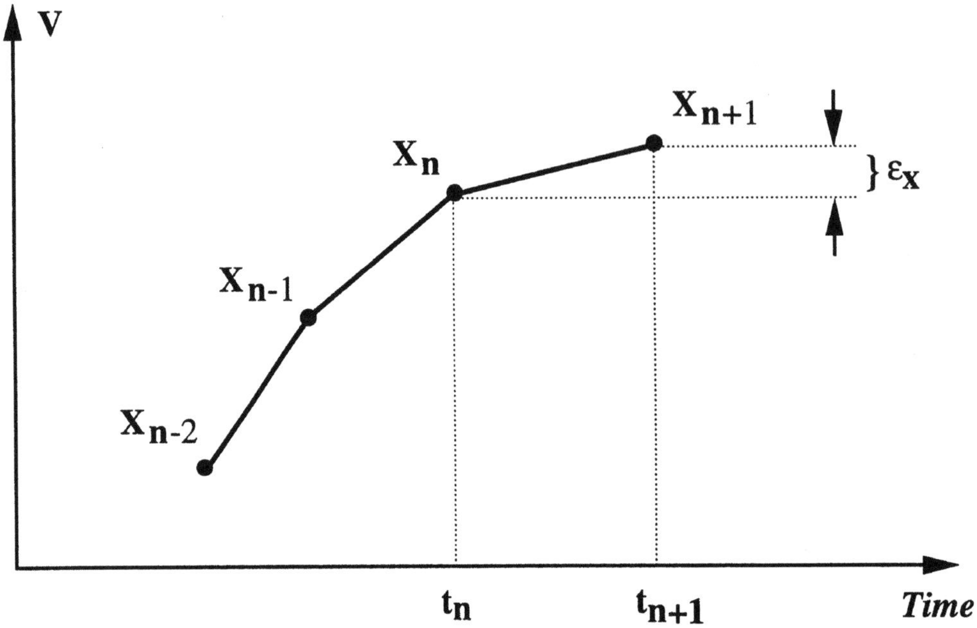

Figure 4.8: Simple Latency Detection

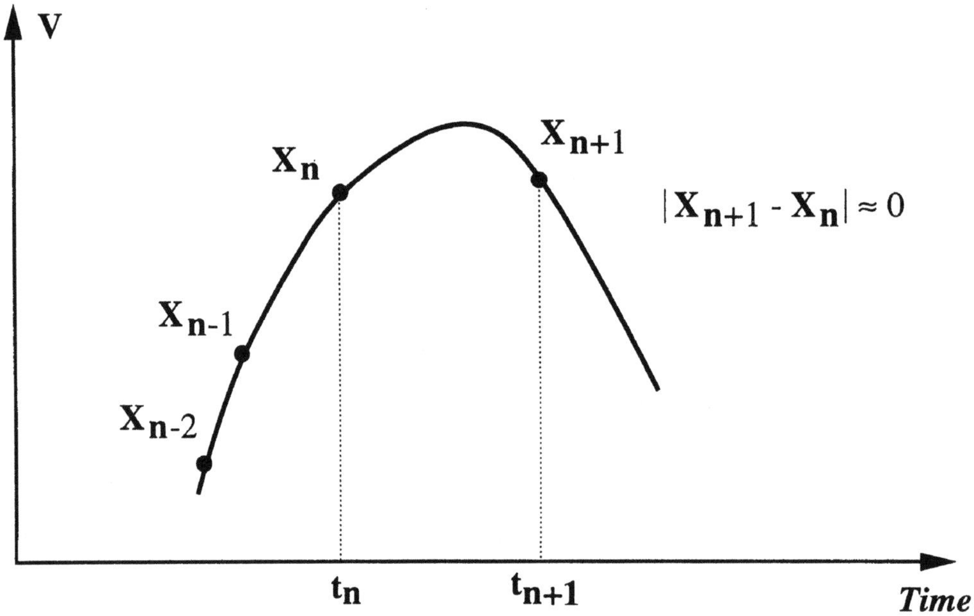

Figure 4.9: Potential Problem in Latency Detection

Latency Condition 1.1:

$$|\mathbf{x}_{n+k} - \mathbf{x}_n| < \varepsilon_x \ , \quad \mathbf{k}>1 \tag{4.7}$$

While this conservative approach works well in practice, it is still not strong enough to handle the general case. For example, if a global variable time-step control is used, the step sizes may be very small due to some fast component resulting in small changes in **x** over a large number of time points (if **x** is a slower component). In this case, it would make more sense to use a rate-of-change criterion to detect latency rather than the absolute change in **x**. That is, use the check

Latency Condition 2:

$$\frac{|\mathbf{x}_{n+1} - \mathbf{x}_n|}{\mathbf{h}_n} < \varepsilon_{\dot{x}} \tag{4.8}$$

As shown in Fig. 4.10, this requires that $\dot{\mathbf{x}} \approx 0$ to satisfy the latency condition. This method also encounters problems with the example in Fig. 4.9 since $\dot{\mathbf{x}} \approx 0$ as the signal switches direction. A more conservative way to do this type of latency check would be to use the strategy of Condition 1.1 and include a number of points from the past.

Latency Condition 2.1:

$$\frac{1}{\mathbf{k}}\sum_{j=1}^{\mathbf{k}} \frac{|\mathbf{x}_{n+2-j} - \mathbf{x}_{n+1-j}|}{\mathbf{h}_{n+1-j}} < \varepsilon_{\dot{x}} \ , \quad \mathbf{k}\geq 1 \tag{4.9}$$

This condition uses an average rate of change based on the previous **k** solutions to detect latency and this overcomes the problem given in Fig. 4.9. However, another problem arises if the true value of $\dot{\mathbf{x}}$ is some small non-zero value that eventually changes the value of **x** significantly at some point in the future. To resolve this problem, a "wake-up" mechanism

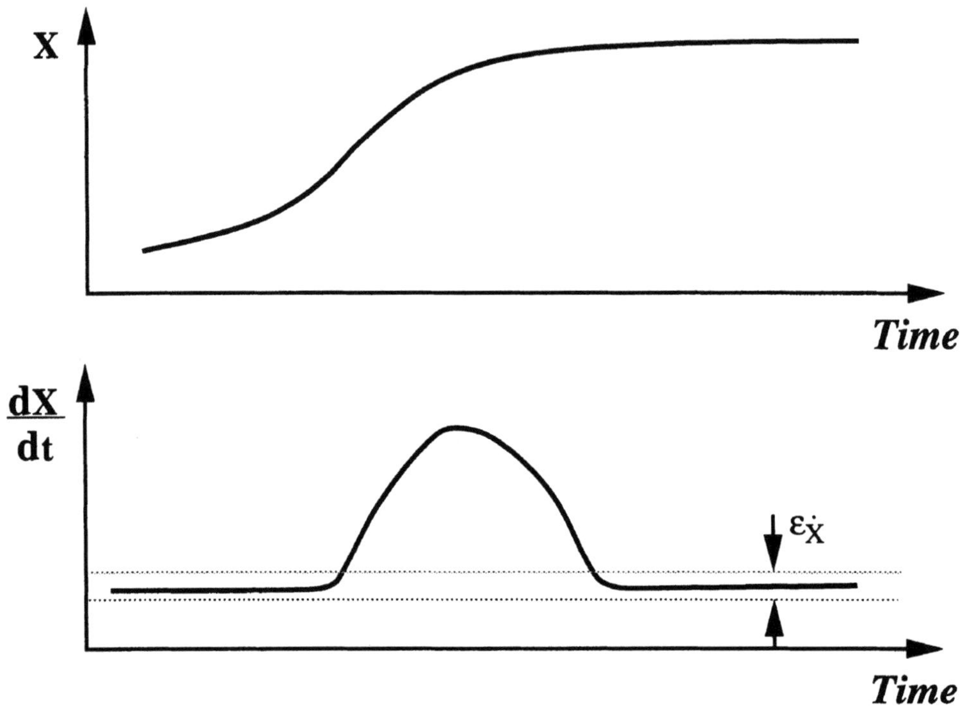

Figure 4.10: Variable Step Latency Criterion Based on Rate-of-Change

should be used with either Condition 1.1 or 2.1 when it is anticipated that component **x** has undergone a significant change in value. That is, the actual rate-of-change of **x** should be used to predict the wake-up time point, as follows:

Wake-up Condition 1:

$$\mathbf{h}_{\mathbf{next}} \frac{|\mathbf{x}_{n+1} - \mathbf{x}_{n}|}{\mathbf{h}_{n}} > \varepsilon_{\mathbf{x}} \tag{4.10}$$

and $\mathbf{t}_{\mathbf{wake-up}} = \mathbf{t}_{n+1} + \mathbf{h}_{\mathbf{next}}$. This wake-up condition can be used to compute $\mathbf{h}_{\mathbf{next}}$ and the component should be re-activated and solved at $\mathbf{t}_{\mathbf{wake-up}}$.

This process is illustrated in Fig. 4.11.

The latency and wake-up conditions specified above work well in practice and their use can be justified by considering latency exploitation as the use of a zeroth-order explicit integration method as described in reference [RAB79]. Explicit integration algorithms are obtained directly from a Taylor series expansion of the solution at the point $\mathbf{t_n}$:

$$\mathbf{x}_{n+1} = \mathbf{x}_n + \mathbf{h}_{n+1}\dot{\mathbf{x}}_n + \frac{\mathbf{h}_{n+1}^2}{2}\frac{\mathbf{d}^2\mathbf{x}_n}{\mathbf{dt}^2} + \cdots \tag{4.11}$$

A zeroth-order method uses only the first term and produces the following trivial integration method for which $\mathbf{x(t_{n+1})}$ is simply updated with the value $\mathbf{x(t_n)}$ at the previous time point:

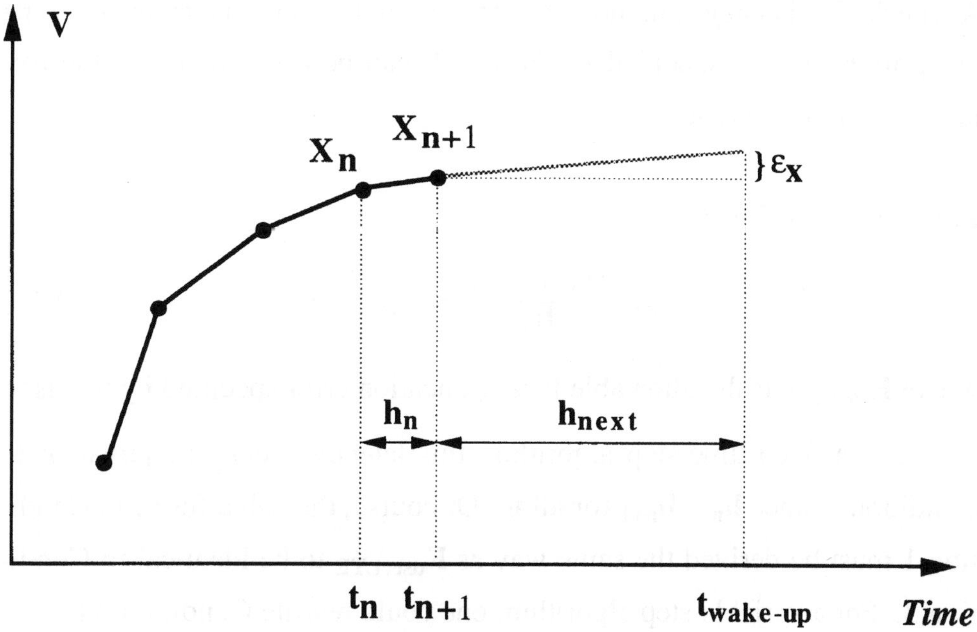

Figure 4.11: Wake-up Mechanism

$$\mathbf{x}_{n+1} = \mathbf{x}_n \tag{4.12}$$

This integration method has a local truncation error (LTE) given by

$$\mathbf{LTE} = \mathbf{h}_{n+1}\dot{\mathbf{x}}(\xi) \qquad \mathbf{t}_n \leq \xi \leq \mathbf{t}_{n+1}$$

An estimate of the LTE can be obtained using a finite difference approximation for $\dot{\mathbf{x}}$:

$$\dot{\mathbf{x}}_n(\xi) \approx \frac{\mathbf{x}_{n+1} - \mathbf{x}_n}{\mathbf{h}_n}$$

Therefore the LTE estimate is given by

$$\mathbf{LTE} \approx \mathbf{h}_{n+1}\frac{\mathbf{x}_{n+1} - \mathbf{x}_n}{\mathbf{h}_n}$$

A check for latency can now be constructed from this analysis. The integration method specified in Eq. (4.12) can be used whenever the following condition is satisfied:

Latency Condition 3:

$$\mathbf{h}_{n+1}\frac{|\mathbf{x}_{n+1} - \mathbf{x}_n|}{\mathbf{h}_n} < \mathbf{E}_{userLTE} \tag{4.13}$$

where $\mathbf{E}_{userLTE}$ is the allowable local truncation error specified by the user.

For a fixed time-step algorithm, this latency check is equivalent to Condition 1 since $\mathbf{h}_n = \mathbf{h}_{n+1}$ for all $\mathbf{n}$. Of course, the value for ε_x in Condition 1 must be derived the same way as $\mathbf{E}_{userLTE}$ to be identical to Condition 3. For a variable step algorithm, one could rewrite Condition 3 as

$$\frac{|\mathbf{x}_{n+1} - \mathbf{x}_n|}{\mathbf{h}_n} < \frac{\mathbf{E}_{userLTE}}{\mathbf{h}_{new}}$$

By replacing $\mathbf{h}_{n+1}$ with a constant value of step size $\mathbf{h}_{max}$ such that

$h_{max} \gg h_{n+1}$, one can provide a somewhat tighter constraint:

$$\frac{|x_{n+1} - x_n|}{h_{n+1}} < \frac{E_{userLTE}}{h_{max}}$$

Then latency condition 2 and 3 can be made identical by setting $\varepsilon_{\dot{x}} = E_{userLTE}/h_{max}$. Note that Condition 3 is an *a posteriori* criterion (i.e., it is used after selecting h_{n+1}) to detect latency. A similar criterion can be used in an *a priori* manner to decide when to activate the component. The idea is to use the LTE requirement to predict the time point when the zeroth-order integration method is no longer valid by checking when Latency Condition 3 is violated:

$$h_{new} \frac{|x_{n+1} - x_n|}{h_n} > E_{userLTE} \tag{4.14}$$

where $h_{new} = t_{wake-up} - t_{n+1}$ and $t_{wake-up}$ is the time when the component should be activated. This wake-up time can be computed as follows:

$$t_{wake-up} = t_{n+1} + \frac{E_{userLTE} h_{n+1}}{x_{n+1} - x_n} \tag{4.15}$$

and this is identical to Wake-up Condition 1. Therefore, the intuitive arguments which lead to Latency Conditions 1 and 2 and Wake-up Condition 1 are well-supported by the above analysis.

4.6.2. Events and Event Scheduling

The next issue is to define precisely the notion of electrical events for use in conjunction with the scheduling algorithm. The proper definition of this concept is important from the standpoint of efficiency and accuracy, as will be seen. In logic analysis, an event occurs when a node makes a transition from one state to another (different) state. The event causes the fanouts of the node to be scheduled in the time queue. As long as the node remains in the same state, no additional events are generated. Since logic states are discrete, logic events are easy to identify. In electrical analysis, there is a continuum of "allowed states" making it more difficult to distinguish a significant event from an insignificant one. However, the definition of logic events can be extended in a straightforward manner to electrical analysis. The resulting definition of an electrical event is connected with the notion of "active" and "latent" components.

Definition 4.2 (Electrical Events)

In electrical analysis, a component is "latent" if it satisfies one of the latency conditions given by Eqs. (4.6-4.9). Otherwise, it is an "active" component making a transition from one electrical value (or state) to another. Active components generate electrical events each time they make a transition to a new value. □

The usefulness of this definition is seen in the following. Consider the two-stage inverter of Fig. 4.12. For this circuit, **A** ∈ **Fanout(I)** and **B** ∈ **Fanout(A)**. As depicted by the arcs in the corresponding graph, there are four ways to schedule nodes:

1) node I can schedule node A (fanout scheduling)
2) node A can schedule node A (self-scheduling)
3) node A can schedule node B (fanout scheduling)
4) node B can schedule node B (self-scheduling).

Whether a given node (say, node A) should actually schedule any events depends on its own state and the state of its fanouts (node B in this case). Since each node can be either "active" or "latent," a total of four cases exist. These cases are listed in Table 4.1 along with the recommended action to be taken by node A for each case.

As the table indicates, case (2) is the only case where the scheduling mechanism is conservative. The other cases do not introduce any additional work or create accuracy problems and therefore are listed as *reasonable*. However, case (2) can be a source of either accuracy problems or

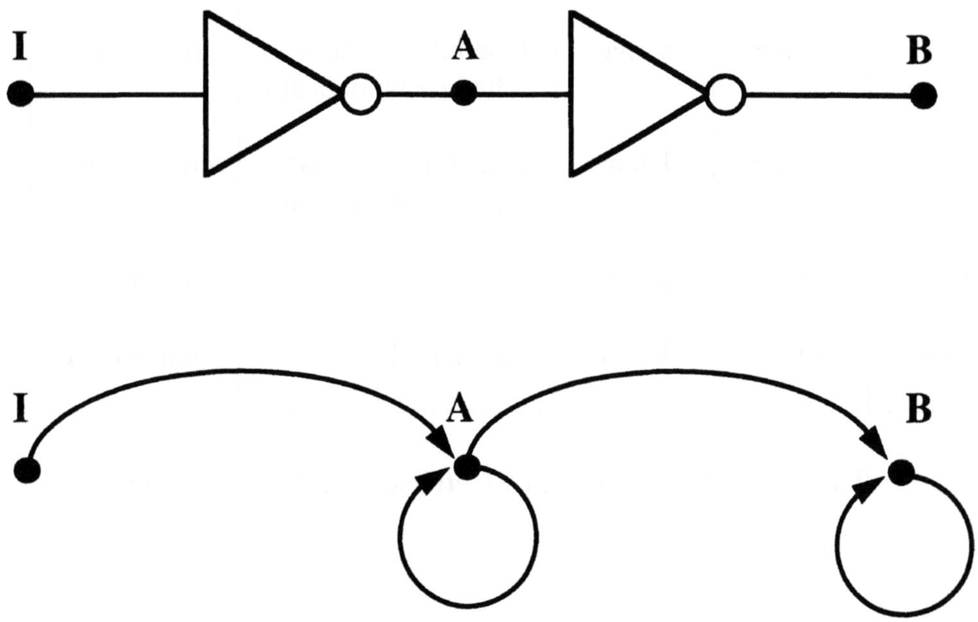

Figure 4.12: Scheduling Possibilities for a Simple Example

excessive computation. To see this, consider the circuit in Fig. 4.13. If node A is "active," it will force nodes B, C and D to be processed if the action recommended in Table 4.1 is taken. In reality, only node B should be processed. The other two nodes do not change due to the bias conditions, but this is not known *a priori*. Therefore, case (2) is considered to be a conservative scheduling strategy. The alternative would be to ask the question: is fanout $\mathbf{x_j}$ *sensitive* to changes in $\mathbf{x_i}$? Here, $\mathbf{x_i}$ = A and $\mathbf{x_j} \in$ **Fanout**($\mathbf{x_i}$) = { B , C , D }. Only an affirmative response to this question causes a particular $\mathbf{x_j}$ to be scheduled by $\mathbf{x_i}$. Otherwise $\mathbf{x_j}$ should not be scheduled.

The conditions associated with case (2) can also be viewed as a wake-up condition due to inputs. That is, "Does the change at node A

case	status of node A	status of node B	action by node A	comment
(1)	active	active	schedule self at t+h schedule fanouts at t	reasonable
(2)	active	latent	schedule self at t+h schedule fanouts at t	conservative
(3)	latent	active	no scheduling req'd	reasonable
(4)	latent	latent	no scheduling req'd	reasonable

Table 4.1: Four Cases in Electrical Event Scheduling

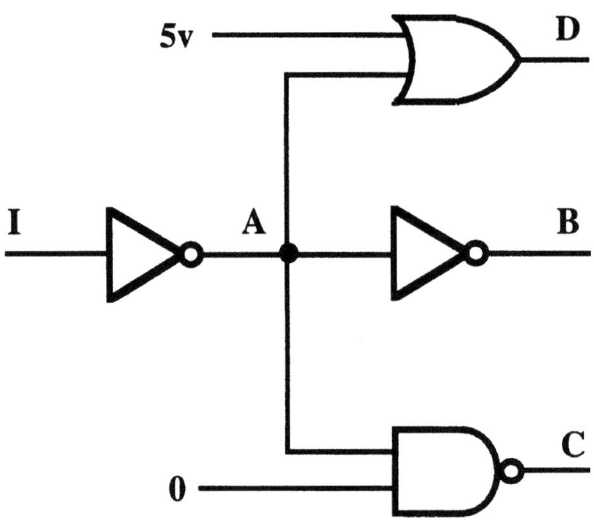

Figure 4.13: Conservative Scheduling Case

wakeup node B?". The previous wake-up conditions were all handled via the self-scheduling mechanism. In this case, the question is whether or not a change at $\mathbf{x_i}$ translates to a change at a fanout $\mathbf{x_j}$ such that $\mathbf{x_j}$ violates its latency condition. Since $\mathbf{x_j}$ may have a number of fanin nodes which are active, superposition must be used to determine the combined effect of all active fanin nodes on $\mathbf{x_j}$. This involves determining the transconductance, $\frac{\partial \mathbf{f_j}}{\partial \mathbf{x_i}}$, and performing the computation:

$$\Delta \mathbf{x} = \frac{\mathbf{h_n}}{\mathbf{C_j}} \sum_{i=1}^{k} \frac{\partial \mathbf{f_j}}{\partial \mathbf{x_i}} \Delta \mathbf{x_i} \tag{4.16}$$

where $\mathbf{k}$ is the number of fanin nodes of $\mathbf{x_j}$ which are active, $\mathbf{h_n}$ is the current step size, and $\mathbf{C_j}$ is the total capacitance at node $\mathbf{x_j}$. This computation assumes that all the additional currents, due to changes in the fanin nodes, charges the capacitances at node $\mathbf{x_j}$. This produces a new wake-up condition due to the inputs, as follows:

Wake-up Condition 2:

$$\mathbf{h_{new}} \frac{|\mathbf{x_{n+1}} - \mathbf{x_n}|}{\mathbf{h_n}} + \Delta \mathbf{x} > \varepsilon_{\mathbf{x}}$$

where $\mathbf{h_{new}} = \mathbf{t_{new}} - \mathbf{t_{latent}}$, and $\mathbf{t_{new}}$ is the current time point. In the worst-case, the computation in Eq. (4.16) can be as expensive as performing an evaluation of $\mathbf{x_j}$, but it certainly is not as accurate. Since there is no way to guarantee that Wake-up Condition 2 is a sufficient check for latency violation, since it is only a local criterion, it is better to perform the evaluation of $\mathbf{x_j}$ rather than the sensitivity check to guarantee that an error is not made inadvertently. This results in a stronger condition for latency, which involves the fanin nodes also being latent.

The ideas presented above are formalized in the following:

1) A component $\mathbf{x_i}$ is defined as being latent if

   a) it satisfies the latency conditions specified in Eqs. (4.6-4.9) and

   b) all $\mathbf{e_k} \in$ **Fanin**($\mathbf{x_i}$) satisfy their latency criteria.

2) A latent component does not generate any events.

3) If a component is not latent, then it is active and hence will generate events for itself and for all $\mathbf{x_j} \in$ **Fanout**($\mathbf{x_i}$) after every transition.

4) A latent component $\mathbf{x_i}$ is scheduled for re-evaluation if

   a) the wake-up condition specified in (4.10) is satisfied, or

   b) any component $\mathbf{e_k} \in$ **Fanin**($\mathbf{x_i}$) becomes active.

4.6.3. Latency in the Iteration Domain

Another form of latency can be exploited at each time point due to the decoupled nature of the relaxation process. Since the components in the system are changing at different rates, it is quite possible that slowly varying components will converge quickly at each time point since their behavior can be predicted accurately. Once these components have converged, there is no need to reprocess them at the same time point unless required to do so by some other component. This form of latency is called iteration domain latency and can also be exploited efficiently using the same event-driven techniques used for time domain latency.

The iteration domain is a discrete space in which a sequence of iteration values of a component can be represented as a function of the iteration number [KLE84]. This iteration domain can be viewed in the same way as the time domain. For example, if a converging sequence of iterations for a component, $\mathbf{x_i}$, is plotted against the iteration number, a waveform is

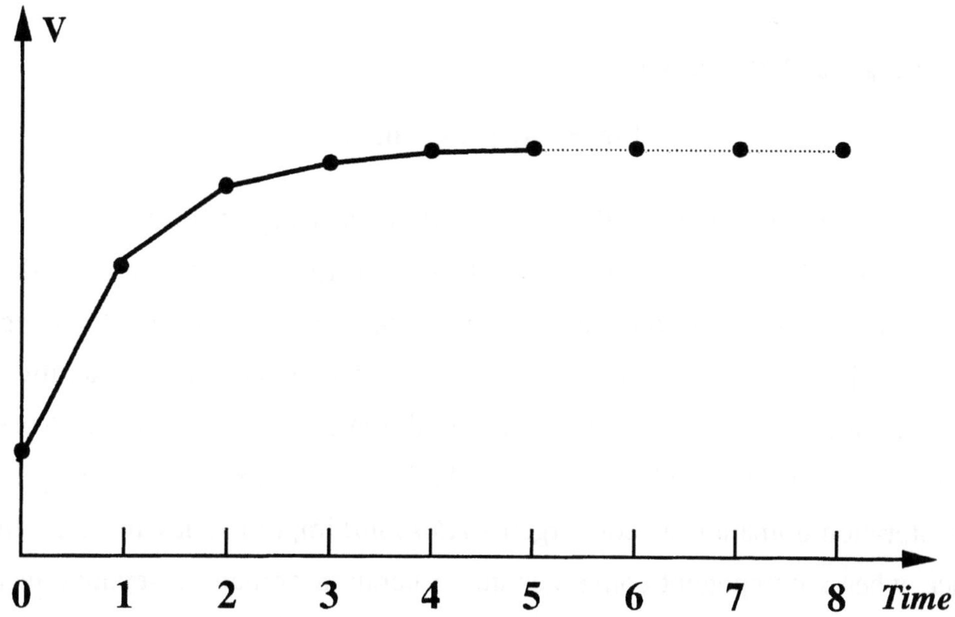

Figure 4.14: Iteration Domain Waveform

produced as shown in Fig. 4.14. The detection of latency in the time domain is seen to be analogous to the detection of convergence in the iteration domain. In fact, since the "step size" is fixed in the iteration domain, the check for convergence should be similar to that for the Latency Condition 1 given earlier. This corresponds to checking if the iteration waveform is "flat enough" [KLE84] and is given as

Convergence Criteria 1:

$$|x_i^{k+1} - x_i^k| < \varepsilon$$

which is consistent with the usual check for convergence. False convergence occurs when the condition is satisfied but the necessary accuracy has not been obtained. Therefore, a check similar to Latency Condition 1.1 would be better to avoid this problem [KLE84].

Convergence Criteria 1.1:

$$|x_i^{k+m} - x_i^k| < \varepsilon, \quad m>1$$

To exploit latency in the iteration domain using event-driven techniques, a table similar to the one for latency in time is necessary. In the iteration domain, if a component is "iterating," it is equivalent to being "active" in the time domain, and if it has "converged" in the iteration domain, it is equivalent to the "latent" condition in the time domain. Note that latency in time implies latency in the iteration domain, but latency in the iteration domain (i.e., convergence) *does not* imply latency in time. In fact, when a component converges in the iteration domain, a separate test is necessary to determine if it is active or latent in the time domain. The four cases in the iteration domain are listed in Table 4.2 below along with the recommended action for node A, assuming that node A is in the "converged" state initially and enters the state listed in column 2 after computing its new value.

Table 4.2 shows that case (2) is again the only conservative scheduling situation. To understand this case, consider Fig. 4.13 again. Each time node A performs an iteration, it will schedule nodes B, C and D. However, as before, only node B should be processed as nodes C and D are latent in time and hence are in the converged state at the time point. If node A requires many iterations to converge, it will schedule nodes C and D many times resulting in a lot of unnecessary work. However, there is no need to repeatedly schedule all its fanouts on every iteration, especially since the nodes have a self-scheduling ability. Therefore, one strategy might be for node A to schedule its fanouts *on every other iteration* rather than on every iteration. This could be used for both case (1) and case (2) since the self-scheduling mechanism would take care of any additional scheduling of node B.

case	new status of node A	status of node B	action by node A	comment
(1)	iterating	iterating	schedule self at t schedule fanouts at t	reasonable
(2)	iterating	converged	schedule self at t schedule fanouts at t	conservative
(3)	converged	iterating	no scheduling req'd	reasonable
(4)	converged	converged	no scheduling req'd	reasonable

Table 4.2: Four Cases in Iteration Domain Latency

CHAPTER 5

GATE-LEVEL SIMULATION

5.1. INTRODUCTION

When the complexity of an integrated circuit design reaches the point where electrical analysis is no longer cost-effective, logic simulation or gate-level simulation may be used. Rather than dealing with voltages and currents at signal nodes, discrete logic *states* are used. In essence, logic analysis may be viewed as a simplification of timing analysis, described in the previous chapter, where the difference equations are replaced by a set of discrete state equations and only simple Boolean operations are required to obtain new logic values at each node. These Boolean operations are generally the most efficient ones available on a digital computer. In a classical logic simulator, transistors are usually grouped into logic *gates* wherever possible and modeled at the *gate-level* rather than at the individual transistor level. This form of simplification, sometimes referred to as *macromodeling*, can result in greatly enhanced execution speed by reducing both the number of models to be processed and simplifying the arithmetic operations required to process each transistor group. With event-driven, selective trace analysis and the above simplifications, asynchronous logic simulators are typically 100 to 1000 times faster than the most efficient forms of electrical analysis.

The major objective of all simulators is to accurately predict the behavior, both normal and abnormal, of the physical circuits they model. This is even more critical in the context of mixed-mode simulation where the overall accuracy may be limited by the accuracy in the higher levels of simulation. Therefore, gate level analysis in a mixed-mode simulator must provide the correct results and at least first-order timing information. The main factors controlling the accuracy of gate level simulation are the state model and the delay model. The delay model must be computationally simple and at the same time include the most important factors contributing to it. Modeling parameters are usually provided with the delay model. If these parameter values are derived from careful characterization of transistor circuits that form the logic gates, then a simplified gate model can be used with a high degree of confidence.

The tradeoff between the accuracy of logic simulation and the computer time required to perform a simulation is very important. For example, the accuracy of logic simulation can be improved by increasing the number of logic states used in the simulation. However, as the number of states increases, the overall runtime may also increase. In fact, the number of logic states, their meaning, the delay models used and the event scheduling algorithm all have a profound impact on the speed and accuracy of logic simulation. The proper choice of each of these factors depends on the circuit technology and its associated characteristics, as well as the particular design methodology used. It is this wide variety of factors that has resulted in the development of a large number of logic simulators, almost every one addressing a different set of tradeoffs.

While it is clear that the transition from the continuous electrical domain to the discrete logic domain may result in the loss of some circuit information, it is important that the circuit design methodology accommodate this type of simplification. Otherwise, the logic simulation mode cannot be used effectively. Unfortunately, in MOS logic circuits, there are many transistor configurations that are not directly amenable to this type of transformation. To overcome this problem, switch-level simulation was developed and has become the preferred form of simulation for MOS logic circuits. This approach is detailed in the next chapter.

In this chapter, some of the factors influencing the choice of logic states and delay models are described. Since logic simulators have been in use for the design of digital hardware since the early 1950s, it is impossible to address all aspects of simulator development here. Therefore, only those aspects which are related to mixed-mode simulation are emphasized. In addition, the modifications necessary to make gate level simulation suitable for the mixed-mode environment are described.

5.2. EVOLUTION OF LOGIC STATES

5.2.1. Two-State Logic Model

The earliest use of logic simulation was for the verification of combinational logic. Since the logic was assumed to have zero delay and logic gates were assumed to implement ideal Boolean operations such as AND, OR and INVERT, only two states were required: a state representing *true* (logic 1) and a state representing *false* (logic 0). With a two-state simulator, it is not only possible to verify the logic function of a digital system (i.e., generate a *truth table*) but it is also possible to detect certain other types of potential design errors such as *hazards* and *races* [EIC65]. A hazard is a momentary incorrect output state, after an input transition, resulting from paths in the circuit with different delay times to the output. There are a number of different types of hazards that can arise in in a logic circuit: static 0 hazard, static 1 hazard, dynamic 1 hazard and dynamic 0 hazard. These hazards are illustrated in Fig. 5.1. A race condition exists in an asynchronous sequential circuit if more than one of the state variables undergoes a transition during a state transient. If the final stable state of the circuit depends on the order in which the state variables change, the race is termed *critical*; otherwise, it is termed *noncritical*.

Although hazards may occur in combinational as well as sequential circuits, they are generally most important when they affect the behavior

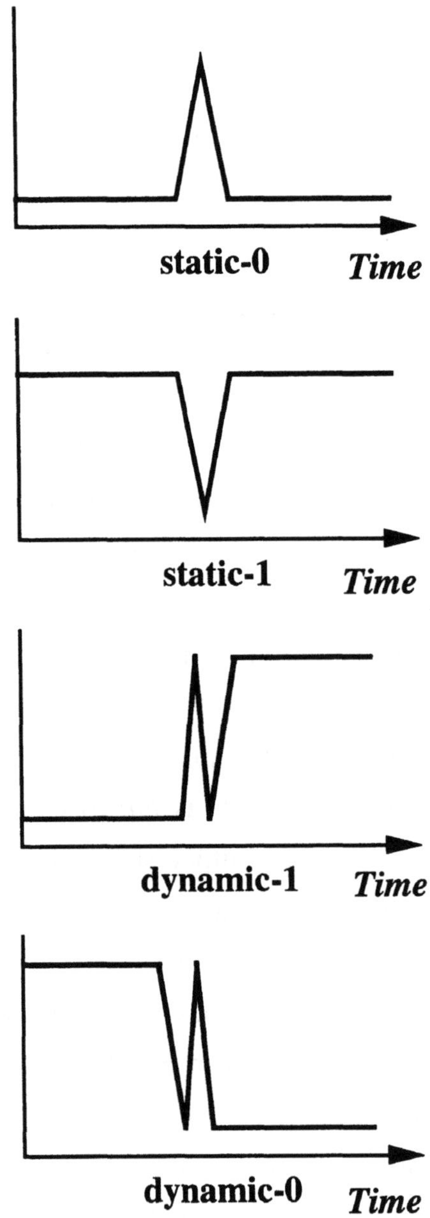

Figure 5.1: Four Types of Hazards in Logic Circuits

of sequential circuits. Since hazards result from paths with different delay times, any hazard *actually causing a circuit to malfunction* will be detected as a critical race or oscillation in the circuit. However, a two-state simulator (even with random delay models) has only a limited capability for detecting races and hazards, if delay variations are not modeled. If several inputs to a logic gate change within a relatively short period of time, it is possible that the order of occurrence of these events may change if gate delays were distributed at slightly different points within their tolerance limits. If the output state of the gate depends on the order in which the inputs change, a potential hazard exists.

It is not sufficient to simply monitor the output of a gate and look for multiple transitions during an input pattern if all *potential* hazards are to be detected. Depending on the order in which the input transitions are processed, the potential hazard may or may not be detected in the zero-delay simulator. This is illustrated in Fig. 5.2 for a simple NAND gate. If input **A** changes first, then output **D** will switch to the 0 state before returning to the 1 state. However, if input **B** changes first, the output will remain at 1 during the input transitions. The potential for both static and dynamic hazards can be detected. However, the errors caused by actual circuit hazards cannot be detected in a two-state simulator without the use of more accurate delay models.

It should be noted that, in a two-state logic system, only one logic gate may drive (or fanin to) any node (often called a *net* in the context of logic design). If more than one gate did fanin to a node, a potential conflict would arise if one gate had a logic 1 at its output and another a logic 0 since it would be unclear what the resulting signal at the node should be. An exceptional case is that of the wired-function (wired-AND, wired-OR), where the node is treated as a logic gate itself and performs a logic function. This is illustrated in Fig. 5.3(a) for an open-collector TTL

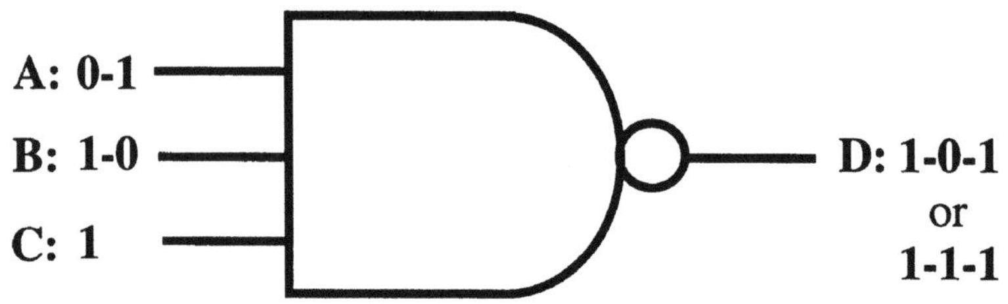

Figure 5.2: Potential Hazard in NAND Gate

example. If it is possible for more than one output to drive a node in a particular technology, such as the so-called *tristate* logic where gates may logically disconnect themselves from the node (as illustrated for MOS in Fig. 5.3(b)), then two-state logic analysis cannot be used to verify the design.

5.2.2. Ternary Logic Model

Two-state simulation has a number of limitations. For example, if two gates drive the same node in the circuit and the output of each gate is different, a conflict situation arises. To model this conflict condition, a third state may be added--the unknown state, **X**. The output node is set to this **X** state whenever any such conflicts arise. The **X** state can then propagate through the fanout gates to other nodes in the circuit and possibly set them to the **X** state. The logic operations for the AND, OR and

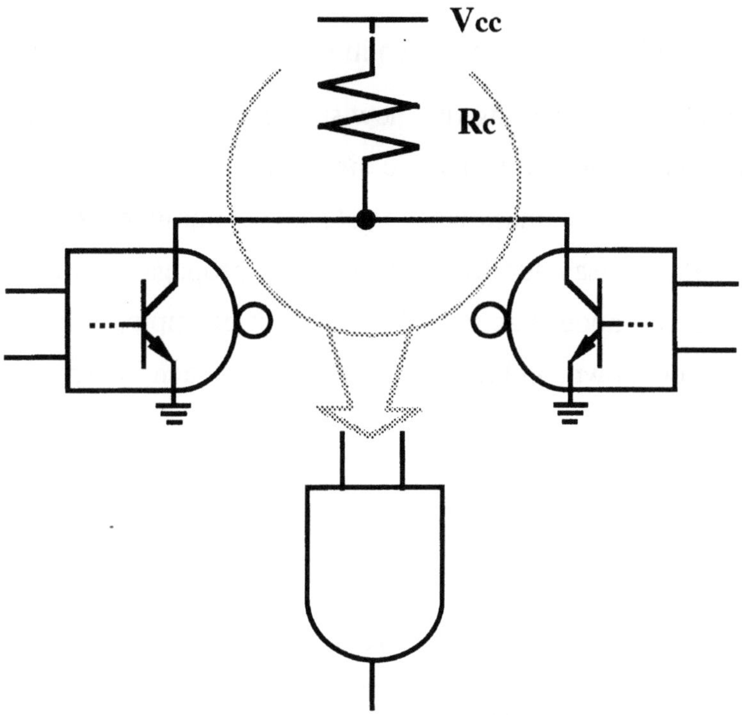

(a)

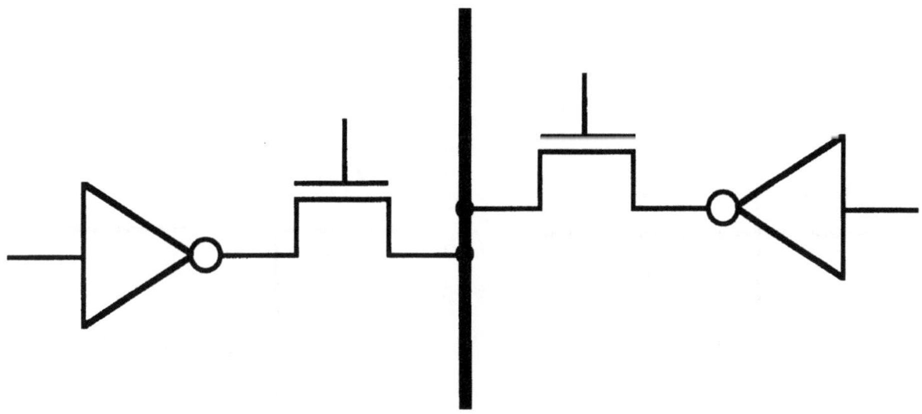

(b)

Figure 5.3: Multiple Devices Driving a Single Node
(a) Open Collector TTL Structure and Its Equivalent Logic Model
(b) MOS Transfer Gates Connected to a Common Bus

INVERT gates with X states are shown in Fig. 5.4

The simple step of adding this new state has caused much confusion and increased the complexity of logic simulation. In [BRE72], the basic problems associated with unknowns in gate-level simulation are described. One such problem arises due to the pessimistic propagation of unknowns when the value of a node is actually known. For example, in Fig. 5.5, one of the inputs is unknown, and this produces an **X** at each intermediate node

a	NOT a
0	1
1	0
X	X

AND	0	X	1
0	0	0	0
X	0	X	X
1	0	X	1

OR	0	X	1
0	0	X	1
X	X	X	1
1	1	1	1

Figure 5.4: Logic Truth Tables Including **X** State

and results in an **X** at the output node **C**. However, since a value of 1 or 0 at that input produces the same results at node **C**, the value at node C is actually known to be 1. Therefore, the propagation of **X** blindly can lead to pessimistic results and excessive computation. This problem can be resolved by keeping track of **X** and $\overline{\mathbf{X}}$ values[1] during the simulation and combining them using the identities $\mathbf{X}\bullet\overline{\mathbf{X}}=0$ and $\mathbf{X}+\overline{\mathbf{X}}=1$ whenever they appear at common AND or OR gate inputs. A second problem with the use of the **X** state is due to the additional complexity it introduces into gate-level logic simulation. In fact, computing the output states of a sequential circuit with **n** inputs and **m** internal states having **k** out of the **n+m** nodes unknown has been shown to be NP-complete with respect to **k** [CHA87].

A number of other problems with the **X** state also exist, mainly due to the misuse of the definition of the state. For example, gate outputs must be correctly initialized prior to the analysis to either the 0 or the 1 state. If a sequential circuit is under analysis, storage nodes such as the output of flip-flops may not be known at initialization time. If the node is set to **X**, there is clearly an inconsistency since the states of **Q** and $\overline{\mathbf{Q}}$ can simply be set to **Q**=1 and $\overline{\mathbf{Q}}$=0 (or equivalently **Q**=0 and $\overline{\mathbf{Q}}$=1) without violating the sanctity of the simulation. Consider the SR flip-flop circuit of Fig. 5.6. If the outputs are assumed to be unknown at initialization, they can not be set to known values due to the input data and the feedback of the **X** states. However, a "conflict" situation does not exist at these output nodes; therefore, the use of **X** in this case is clearly incorrect. Another state is required to account for uninitialized nodes in sequential circuits. A distinction should be made between *initial* unknowns $\mathbf{X_i}$ and *generated* unknowns $\mathbf{X_g}$. When an initial unknown is encountered during the simulation, it can be

[1] Multiple **X** and $\overline{\mathbf{X}}$ states must be maintained, one for each different source of the **X** state.

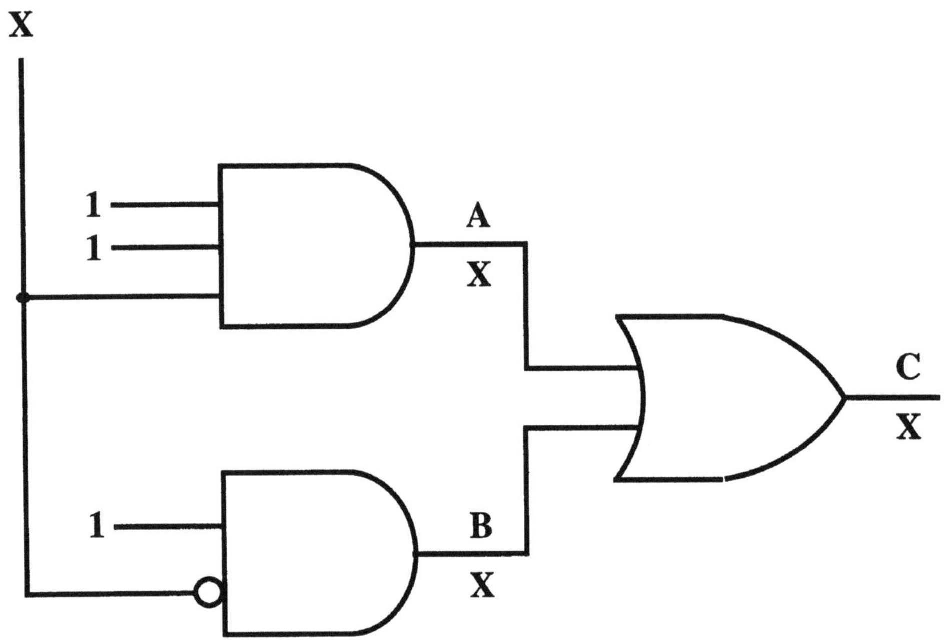

Figure 5.5: Problem Using **X**-State in Gate-Level Simulation

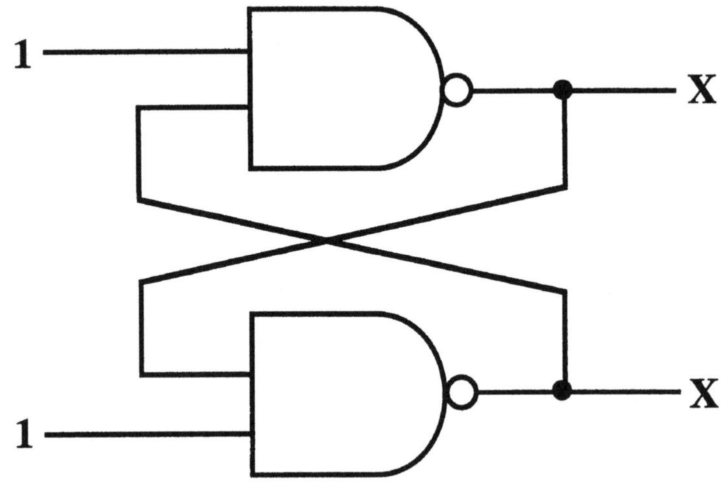

Figure 5.6: Initial Unknowns in a SR Flip-Flop

set to a known value in the processing of the gate it is driving. If a generated unknown is encountered, it must not be set but rather propagated through the gates. The difference between initial and generated unknown states can also prove useful in determining those parts of a circuit not exercised during the simulation (still at $\mathbf{X_i}$ after the simulation).

The **X** state has also been used occasionally for the transition period between 0 and 1, which is another improper use of **X**. For this situation, a **T** state (i.e., transition state) should be employed, or possibly the **R** (rising) and **F** (falling) states to provide information about the direction of the signal transition. In mixed-mode simulation, the use of **X** is generally not recommended. However, the **R** and **F** states are extremely useful and an important part of electrically-oriented gate-level simulation, as described later in this chapter.

5.2.3. A Four-State Logic Model

The ternary logic model described above is still not sufficient for the analysis of general MOS digital circuits which contain transfer gates and tri-state logic circuits. For these circuits, many gate outputs may be connected to a single node, as shown in Fig. 5.7, and it is necessary to determine which output is controlling the state of the node, or *bus*. If more than one gate is forcing the node, a *bus contention* warning must be generated by the simulator. It is possible to represent the condition where the output of M1 is not controlling the bus (G1 is logic 0) by setting the output of M1 to **X** in that case. If this technique is used, there is no longer any distinction between the true unknown state and the *off* condition of the gate. With the addition of a fourth static state, *high impedance* (**Z**) or *non-forcing*, the distinction is maintained. The four static states are illustrated in Table 5.1. A high voltage is represented by logic 1, low voltage logic 0, and unknown is **X**. The fourth state, **Z**, is shown separately since it does

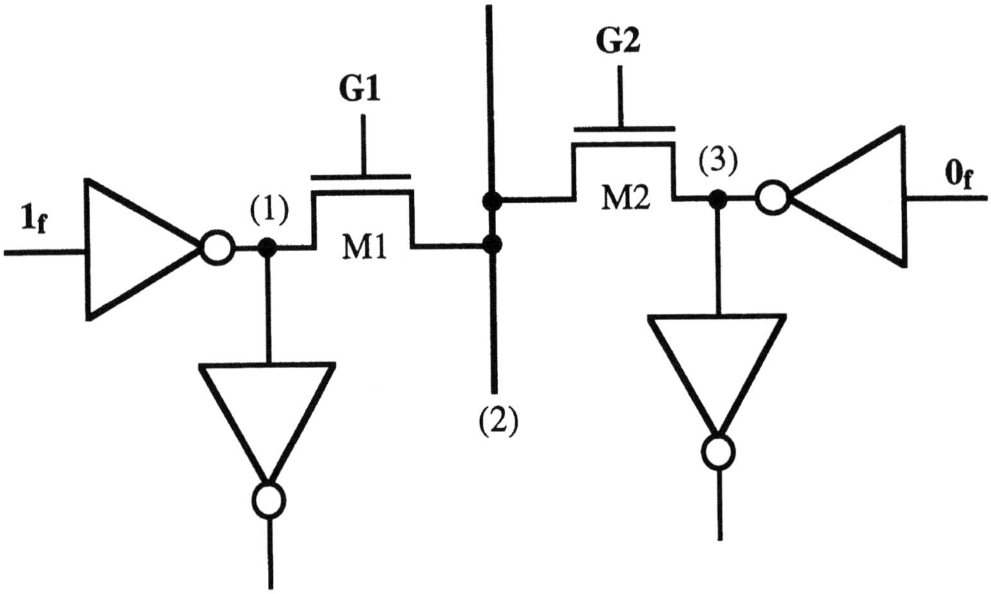

Figure 5.7: Multiple Transfer Gates at a Common Bus

Z		
0	**X**	1

Table 5.1: Four-State Logic Simulation

not represent a voltage state but rather an impedance condition. With the addition of the **Z** state, bus contention can be predicted without confusion. But what if all the gates driving a node are off? What is the input to a fanout gate in this case? In MOS circuits, the previous output is generally "stored" on the parasitic capacitance at the node and held for some time.

This may be modeled by saving two states at each node, the present state and the previous state[2]. If the present state is **Z**, then the previous state can be used to determine the input to fanout gates.

5.2.4. A Nine-State Logic Model

Another approach that can be used to keep track of the previous state of high-impedance nodes is to add three new static states, as shown in Table 5.2. The low impedance states are called *forcing* states (0_f, $\mathbf{X_f}$, 1_f), and there are now three high impedance states (0_z, $\mathbf{X_z}$, and 1_z), which also carry the information about the previous signal level.

0_z	$\mathbf{X_z}$	1_z
0_f	$\mathbf{X_f}$	1_f

Table 5.2: Six-State Logic Simulation

Consider once again the circuit of Fig. 5.7. If M1 and M2 are both conducting, it is clear that the state at node (2) can be determined from our simple model. But what about nodes (1) and (3)? Since the transfer gates are bidirectional, the signal at node (2) may force nodes (1) and (3) to the **X** state. In reality, the output impedance of the inverter is probably considerably lower than the output impedance of the transfer gate and, hence, the inverter output would determine the node state. To model this effect, another three states may be added, called *soft* states, (0_s, $\mathbf{X_s}$, and 1_s), which correspond to the output of the transfer gate when its gate node is on and its input is a forcing or soft state. These states are shown in Table 5.3.

[2] The previous state is required to accurately model storage elements in any case.

Conceptually, the y-axis of this state table may be considered an impedance axis and the x-axis as a voltage axis. In fact, the output of any logic gate may be mapped into this state table by measuring its output voltage (or current) and output impedance. As will be seen later, this technique may also be used to translate gate outputs from logic analysis into electrical inputs for mixed-mode analysis.

0_z	X_z	1_z
0_s	X_s	1_s
0_f	X_f	1_f

Table 5.3: Nine-State Logic Simulation

5.3. CHARACTERIZATION OF SWITCHING PROPERTIES

One aspect of logic simulation that takes on greater significance in the context of mixed-mode simulation is the representation of logic waveforms. In standard logic simulation, the waveforms are represented using the symbols "1" and "0" for the high and low values, respectively, and logic transitions are represented as ideal steps. The rise and fall transition times of the waveforms in standard logic simulation are not as important as the propagation delays from the input to the output of a gate. However, this is not the case in mixed-mode simulation. The transient characteristics during switching are much more important than the propagation delay. If needed, the propagation delay can always be derived from measurements on the waveforms for the input and output nodes.

It is important to have finite nonzero rise and fall delays in the mixed-mode environment for two reasons. First, from a practical

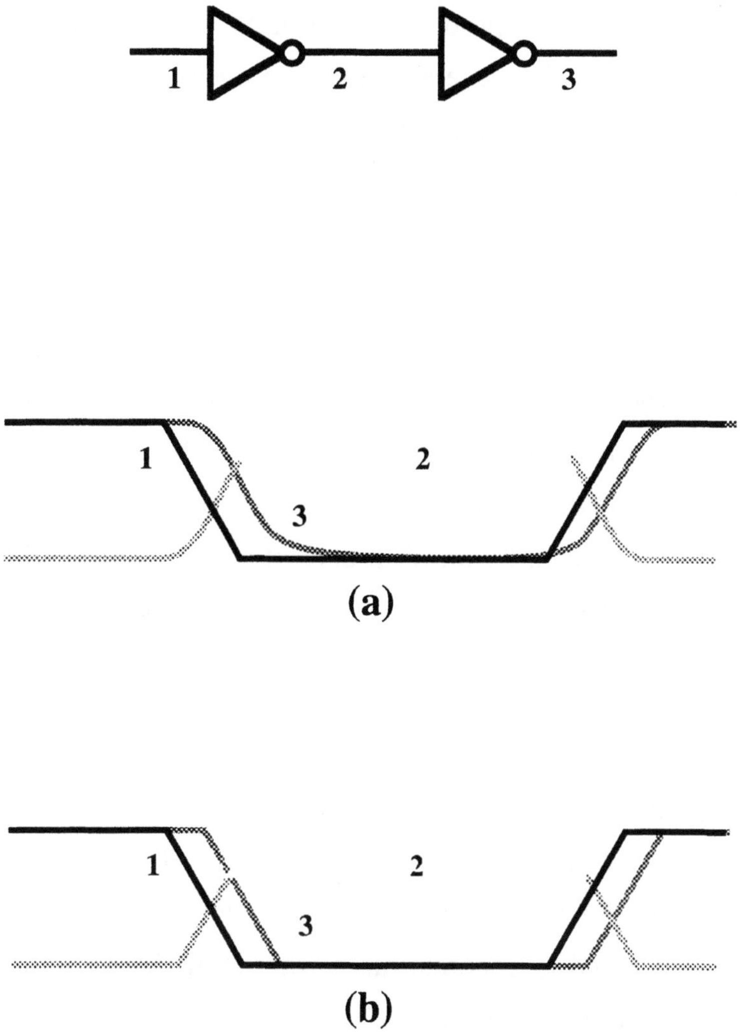

Figure 5.8: (a) Two Inverters (b) Actual Waveforms for inverter chain (c) logic Waveforms for inverter chain

viewpoint, this is not a realistic situation. The capacitance associated with each node produces some finite delay for both rising and falling signals. Second, it will undoubtedly cause convergence problems in the electrical simulation algorithms, specifically in the Newton method, due to abrupt changes in the logic waveforms that feed the electrical portions of the circuit. Therefore, the goal of logic analysis in the context of mixed-mode simulation should be to produce waveforms that are similar to the waveforms that would be generated by pure electrical simulation of the same circuit, albeit with less precision.

A modeling technique that satisfies this requirement can be developed by examining the electrical properties of gates. In Fig. 5.8(a), the output waveforms for a chain of two inverters are shown. The waveforms are characterized by three regions: a region where the output is low, a region where the output is high and a region where the output is in transition. A first-order model of the charging and discharging behaviors at each node is shown in Figs. 5.9(a) and 5.9(b), respectively. In both cases, the model is given by an ideal current source connected to a linear capacitor. The response at the output node is a ramp function that is either rising or falling at a rate that depends on the value of the capacitance and current. In reality, the charging or discharging current is not constant so, for a first-order model, an average current must be used to obtain the approximate timing information. In addition, the capacitance is not constant but an average can also be used for it. The logic waveforms corresponding to the circuits in Fig. 5.9 are shown in Fig. 5.8(b). This approach can be used to generate ramp waveforms for logic gate outputs by simply computing the rise and fall transition times. The details of the transition time computation are left to the next section.

The three regions described above can be represented by four parameters: a low level, a high level, a low threshold and a high threshold. These regions and parameters are shown in Fig. 5.10(a). The four parameters have a direct correlation with the parameters that represent the dc voltage transfer characteristic (VTC) for a logic gate as shown in Fig. 5.10(b).

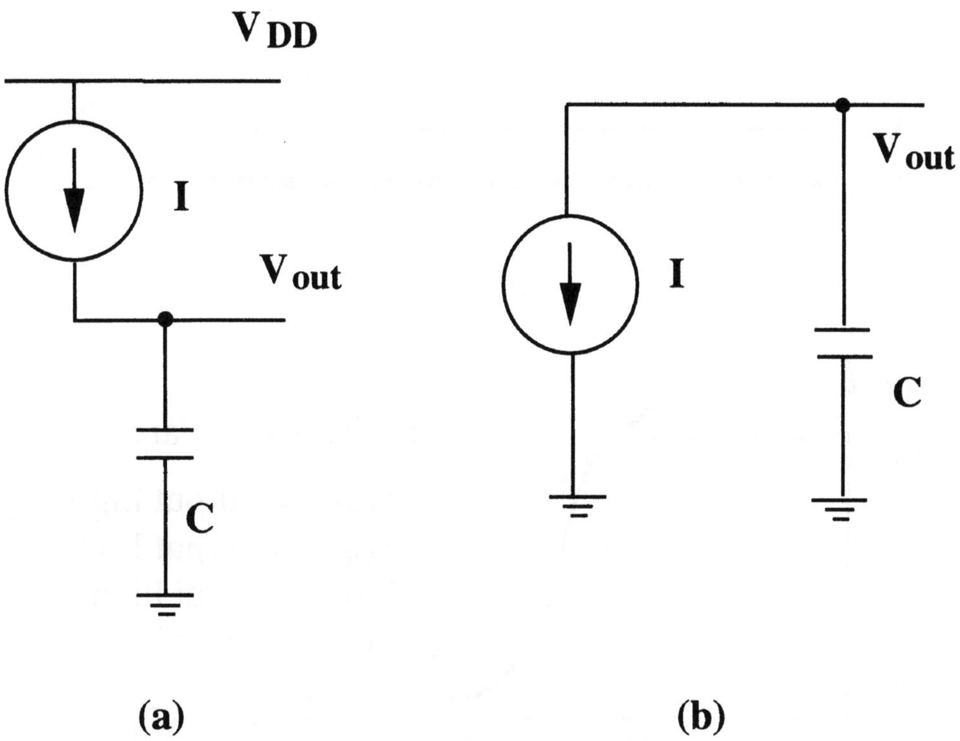

Figure 5.9: (a) First-Order Charging Model
(b) First-Order Discharging Model

This is a graph of the output voltage, $\mathbf{V_{out}}$, versus the input voltage, $\mathbf{V_{in}}$, for a simple inverter. The four parameters in the figure are as follows:

$\mathbf{V_{OL}}$ = low output of inverter
$\mathbf{V_{OH}}$ = high output of inverter
$\mathbf{V_{IL}}$ = maximum value of input before output begins
to drop appreciably

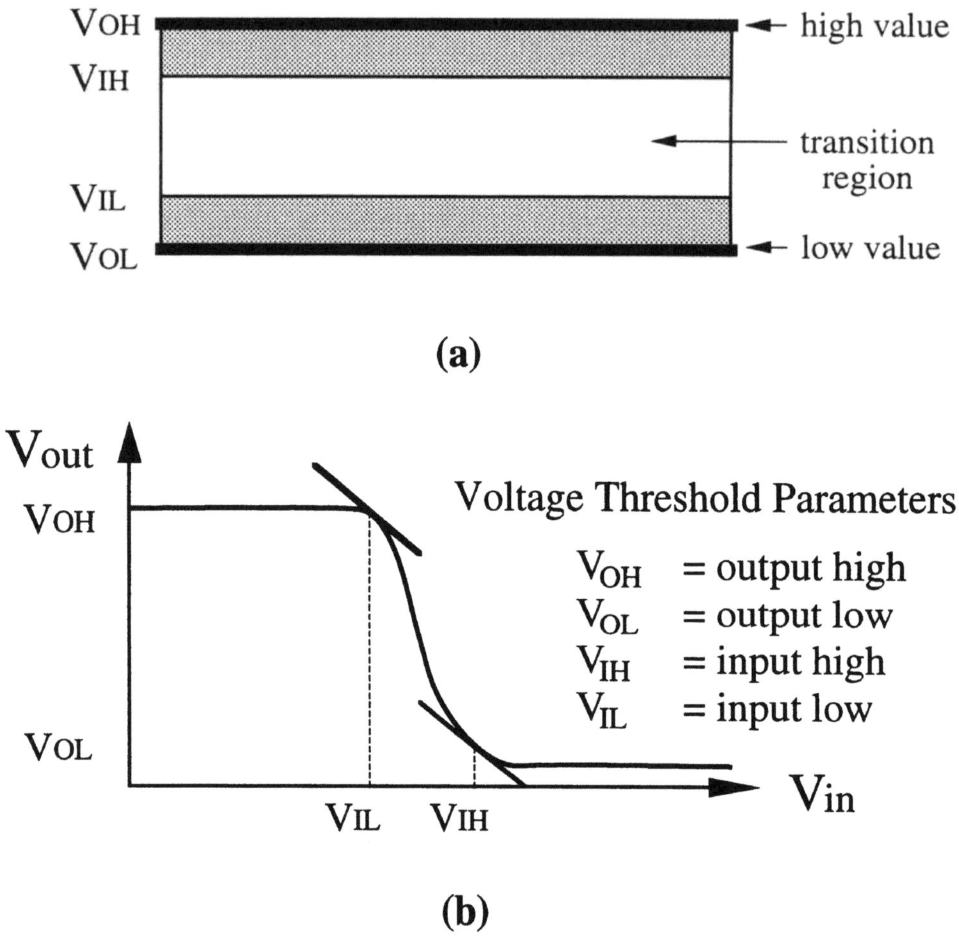

Figure 5.10: (a) Switching Regions
(b) Inverter Voltage Transfer Characteristic

V_{IH} = minimum value of input before output begins to rise appreciably

These parameters are used to define the logic noise margins for the inverter:

$$\mathbf{NM_H = V_{OH} - V_{IH}}$$

$$\mathbf{NM_L = V_{IL} - V_{OL}}$$

The values of $\mathbf{V_{IH}}$ and $\mathbf{V_{IL}}$ based on the definitions above are somewhat arbitrary. Physically, $\mathbf{V_{IL}}$ is the largest value of input voltage that still maintains a valid high voltage at the output, and $\mathbf{V_{IH}}$ is the smallest value of input voltage that maintains a valid low output voltage. A more precise definition can be obtained by examining the input and output relationships. Clearly, the output voltage is some function of the input voltage:

$$\mathbf{V_{out} = f(V_{in})}$$

If some voltage noise, $\mathbf{V_{noise}}$, is superimposed on the input, then

$$\mathbf{V_{out}^{new} = f(V_{in}+V_{noise})}$$

If the right-hand side is expanded in a Taylor series, then the following is obtained:

$$\mathbf{V_{out}^{new} = f(V_{in}) + \frac{\partial f(V_{in})}{\partial V_{in}} V_{noise} + higher\text{–}order\ terms}$$

Therefore,

$$\mathbf{V_{out}^{new} = V_{out}^{old} + gain \times V_{noise} + higher\text{–}order\ terms}$$

From this equation, assuming that the higher-order terms are negligible, it is seen that if the gain is small, the noise is attenuated. However, if the gain is large, the noise is amplified and added to the output. A reasonable breakpoint between the two cases occurs when the gain is 1. Therefore, useful definitions for both $\mathbf{V_{IL}}$ and $\mathbf{V_{IH}}$ are the points along the VTC where

$$|\partial V_{OUT}/\partial V_{IN}| = 1.$$

Although, in reality, the output begins to change before these two critical points are reached at the input, an ideal logic model assumes that no change will occur at the output until the thresholds are exceeded.

In terms of a logic state model, a new four-state logic model [SAK81] is needed, where the state, **s(t)**, at any node at time **t** is an element of the set { 0, **R**, **F**, 1}, where **R**=rising waveform and **F**=falling waveform. Clearly, each of the states, **s(t)**, may be defined in terms of the corresponding node voltages, **v(t)**, and the following noise margin parameters:

$$\mathbf{s(t) = 0 \text{ iff } v(t) \in [V_{OL}, V_{IL})}$$

$$\mathbf{s(t) = 1 \text{ iff } v(t) \in (V_{IH}, V_{OH}]}$$

$$\mathbf{s(t) = R \text{ iff } v(t) \in [V_{IL}, V_{IH}]} \text{ and } \mathbf{\dot{v}(t) > 0}$$

$$\mathbf{s(t) = F \text{ iff } v(t) \in [V_{IL}, V_{IH}]} \text{ and } \mathbf{\dot{v}(t) < 0}$$

The four-state logic model can be represented in truth table form for the AND, OR and INVERT gates as shown in Fig. 5.11. However, the actual transitions from one state to another are governed by practical considerations. Specifically, the transitions 0→**R**, **R**→1, 1→**F**, **F**→0, **R**→**F** and **F**→**R** are permitted. These legal state transitions can be defined in terms of a state diagram as shown in Fig. 5.12. The transitions 0→1, 1→0, 1→**R** and 0→**F** are considered to be illegal since it is physically impossible to make these transitions without either visiting the intermediate states or violating the voltage limits of the circuits.

Encountering an illegal state transition during the simulation is an indication that a timing error may be present in the circuit. As an

a	NOT a
L	H
F	R
R	F
H	L

AND	L	F	R	H
L	L	L	L	L
F	L	F	F	F
R	L	F	R	R
H	L	F	R	H

OR	L	F	R	H
L	L	F	R	H
F	F	F	R	H
R	R	R	R	H
H	H	H	H	H

Figure 5.11: Truth Tables for Four-State Logic Model

example, consider the AND gate in Fig. 5.13(a). The transitions at the inputs and outputs is specified using a string of values that indicate the state of the node in each time slot. If the two input transitions are separated in time, as in Fig. 5.13(b), there is no transition at the output. However, if the input transitions overlap, then the output may attempt an illegal state transition, which indicates that a race condition exists at the input. If input B makes the first transition but the two input transitions still overlap, the output will be a glitch but will not encounter any illegal states as shown in Fig. 5.13(c). Therefore, the output state transitions will either imply an error explicitly or implicitly, but in both cases a timing error can be uncovered.

5.4. LOGIC SWITCHING DELAY MODELS

Now that an appropriate logic transition model has been defined, the next step is to specify the details of the delay calculations. A variety of different delay models have been used in logic simulators and they have evolved over time due to changes in technology in much the same way as the logic model. The simplest delay model is the *zero-delay* model mentioned earlier. This type of model only allows for functional verification

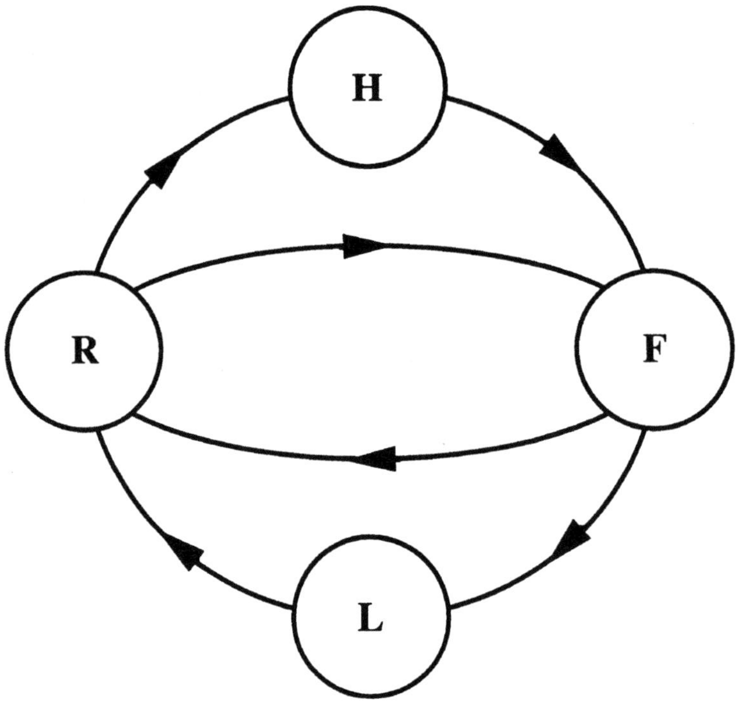

Figure 5.12: Four-State Logic Transition Model

(a)

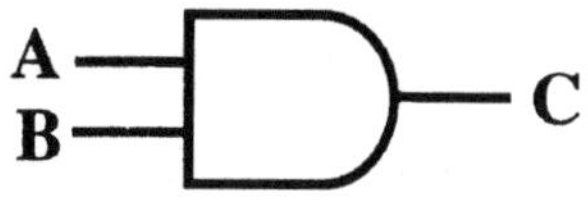


```
(b) A:11111FFFFFF000000000
    B:00000000000RRRRR1111
    C:00000000000000000000  (no change)

(c) A:11111FFFFFF000000000
    B:00000000RRRRR1111111
    C:00000000FFF000000000  (illegal transition: race)

(d) A:11111111FFFFF0000000
    B:0000RRRRRR1111111111
    C:0000RRRRFFFFF0000000  (legal transition: glitch)
```

Figure 5.13: Potential Timing Errors Due to Input Variations

of logic circuits but does not allow the detection of races or hazards and, of course, it does not provide any timing information. It is also prone to problems such as "infinite looping" if there is an odd number of signal inversions in any logic feedback path. Early logic simulators used *unit delay* models to represent timing. In a unit delay simulator, all gates have the same (unit) delay between signal transitions. For logic circuits constructed from a single gate type having similar rise and fall delays, the unit delay simulator provides useful information and lends itself to efficient implementation. If more than one gate type is used, *assignable delays* can provide improved accuracy in the results. In the assignable delay simulator, the delay of the logic gates may be assigned an integer value, $\mathbf{T_D}$. This delay is a multiple of some fundamental analysis time-step, or *minimum resolvable time* (mrt). Here, the **mrt** is the minimum non-zero delay of a logic gate and its value depends on the technology being simulated. For example, the **mrt** may be 1 ns for NMOS, while a value of 100 ps may be appropriate for ECL circuits.

There are two ways in which gate delays may be interpreted as illustrated in Fig. 5.14. A *transmission line* or group delay model propagates

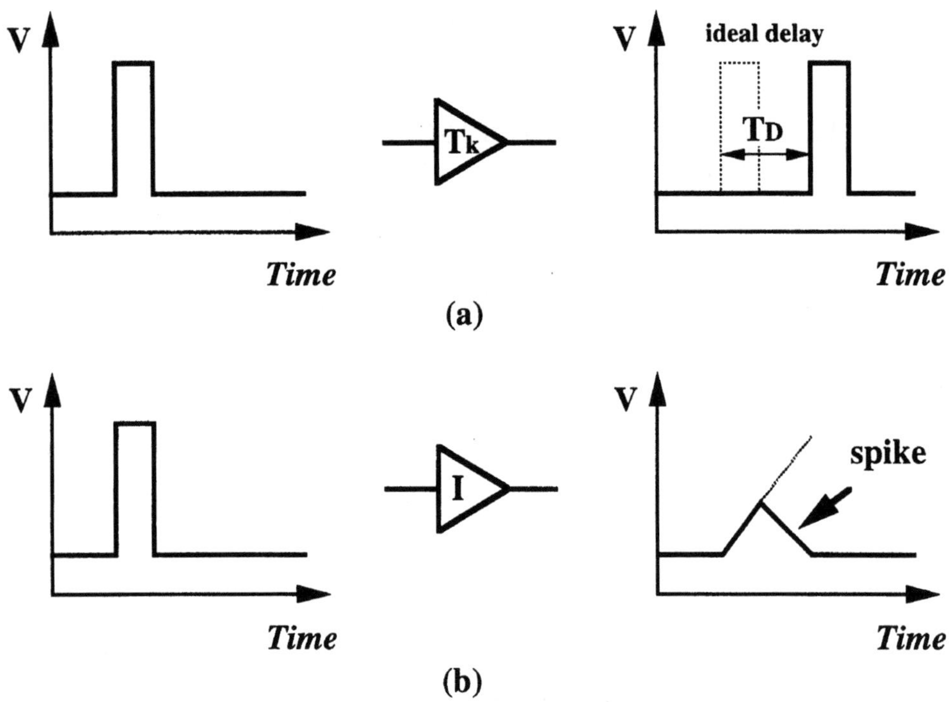

Figure 5.14: Interpretation of Gate Delay
(a) Transmission line model (b) Inertial delay model

the input patterns directly to the output, delayed by some amount $\mathbf{T_D}$. Even very short pulses are propagated unaltered, as shown in Fig. 5.14(a). A second approach is to use an *inertial delay* model, in which the "inertia" or response time of the gate is modeled. If an input event occurs in less time than the time required for the gate to respond, it will be lost as shown in Fig. 5.14(b). Note that in this case *spikes* or *glitches* may be generated at the output. A spike is defined as a signal of shorter duration than necessary to change the state of an element. Spikes may be generated by input hazards or by very narrow input pulses to a gate. A spike may be propagated into the fanout gates as either a new state (E for "error condition") or it may be deleted from the output and a warning message printed. The latter technique generally provides more information from the analysis since the spike is generally an error and will be removed by the designer. By not propagating the spike, more information may be obtained about the correct operation of the circuitry.

For mixed-mode simulation, the delay model used for the switching behavior must be derived from the electrical characteristics. The delay calculation should be based on a transition delay because of the nature of the logic model described in the previous section. For logic circuits in which rise and fall delays vary widely (such as single channel MOS), it is necessary to provide both rise ($\mathbf{t_{LH}}$) and fall ($\mathbf{t_{HL}}$) transition delays for each gate. These delays are a function of a number of different parameters. In MOS circuits, the switching time may depend on

1) the device sizes

2) the supply voltage

3) the output capacitance

4) the number of inputs to the gate, and which one switches in value

5) and the shape (rise or fall times) of input waveforms.

Very few logic simulators have actually incorporated all of the above factors into the delay calculation. However, it is essential that an electrically-oriented logic simulator include the important first-order effects in the delay equation. To derive such an equation, consider the rise and fall delays of the CMOS inverter shown in Fig. 5.15. The fall time, t_{HL}, is given by [UYE88]:

$$t_{HL} = \frac{C_{out}}{\beta_N(V_1 - V_{TN})}\left\{\frac{2V_{TN}}{V_1 - V_{TN}} + \ln\left[\frac{2(V_1 - V_{TN})}{V_0} - 1\right]\right\}$$

and the rise time, t_{LH}, is given by [UYE88]:

$$t_{LH} = \frac{C_{out}}{\beta_P(V_1 - |V_{TP}|)}\left\{\frac{2|V_{TP}|}{(V_1 - |V_{TP}|)} + \ln\left[\frac{2(V_1 - |V_{TP}|)}{V_0} - 1\right]\right\}$$

where C_{out} is the loading capacitance, V_{TN} is the n-channel threshold voltage, V_{TP} is the p-channel threshold voltage, and V_1 =

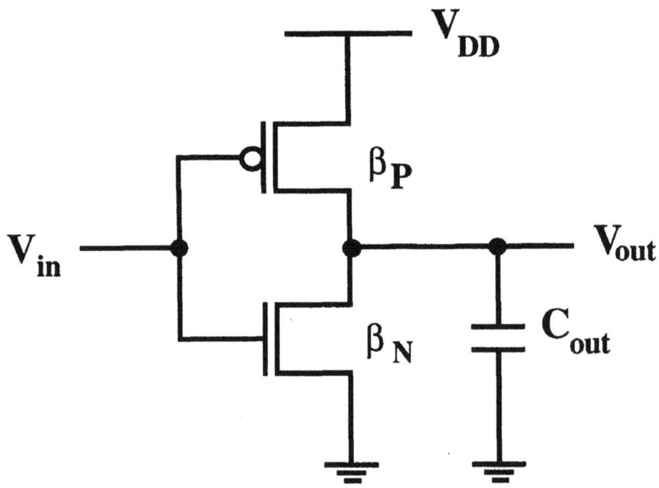

Figure 5.15: CMOS Inverter

$\mathbf{V_{OH}} - 0.1(\mathbf{V_{OH}} - \mathbf{V_{OL}})$ and $\mathbf{V_0} = \mathbf{V_{OL}}+0.1(\mathbf{V_{OH}} - \mathbf{V_{OL}})$ are the 90% and 10% switching points, respectively. All of these parameters are constant except for the output loading capacitance which depends on the number of fanouts connected to the output node.

The delay can be separated into two components by dividing $\mathbf{C_{out}}$ into $\mathbf{C_{intrinsic}} + \mathbf{C_{fanout}}$, where $\mathbf{C_{intrinsic}}$ is the unloaded output capacitance and $\mathbf{C_{fanout}}$ is due only to external gates connected to the node. Then, the total gate delay can be represented by four parameters : the intrinsic gate delays (**tr**, **tf**) and the gate drive-capabilities (**trc**, **tfc**), where

tr = rise time for unloaded gate (y-intercept)
tf = fall time for unloaded gate (y-intercept)
trc = gate drive-capability for rising signals (slope)
tfc = gate drive-capability for falling signals (slope).

Using these values, the total delays are calculated with the equations:

$$\mathbf{t_{LH}} = \mathbf{tr} + \mathbf{trc}^*\mathbf{C_{fanout}} \tag{5.1a}$$

$$\mathbf{t_{HL}} = \mathbf{tf} + \mathbf{tfc}^*\mathbf{C_{fanout}} \tag{5.1b}$$

The logic gates can be characterized to determine the four parameters (**tr**, **tf**, **trc**, **tfc**). The value of $\mathbf{C_{fanout}}$ requires that the input and output capacitances be specified for each gate. For example, in MOS circuits, **ci** can be defined as the MOS capacitance associated with the gate input and **co** can be defined as the drain/source and wiring capacitance. Then the total capacitance at each node becomes

$$\mathbf{C_{fanout}} = \mathbf{co} + \sum_{k=1}^{n} \mathbf{c_{i,k}}$$

This process is shown in Fig. 5.16. Often the delay is a function of the input slope, $\mathbf{s_i}$. This aspect can be incorporated into the premultipliers **trc** and **tfc**:

$$\mathbf{t_{LH} = tr + trc(s_i)*C_{fanout}} \tag{5.2a}$$

$$\mathbf{t_{HL} = tf + tfc(s_i)*C_{fanout}} \tag{5.2b}$$

It is now possible to model the delay by generating a set of curves similar to Fig. 5.17 for every primitive element (NANDs, NORs, inverters, etc.) using accurate electrical simulation. In this figure, the rise and fall times are plotted as a function of the output capacitance. A step voltage is assumed at the input of each gate. Although not strictly true, the relationship between the capacitance and delay is usually taken to be linear. That is, the delay is calculated based on the model of a constant current source charging a linear capacitor. The y-intercept of each curve represents the intrinsic unloaded rise/fall delay while the slope of each curve represents the gate "pull-up" or "pull-down" resistance. Separate characteristics are required for rising and falling outputs if the delay times are not symmetric.

5.5. LOGIC SIMULATION ALGORITHM

The pseudo-code in Algorithm 5.1 provides a simplified description of the logic simulation algorithm based on the previous sections. First, the event time $\mathbf{t_n}$ is established by *NextEventTime()*. All input sources, $\mathbf{e_k}$, that are changing at that time schedule their immediate fanouts. Then all the nodes scheduled at $\mathbf{t_n}$ are processed in sequence until the events at that time are exhausted.

In the algorithm, a node is processed by computing the output value of its associated gate using the states of the inputs at $\mathbf{t_n}$. The input state consists of a voltage value and information indicating whether the signal is rising, falling or stationary. If the new output state is different from the old one, a delay calculation is performed. If the event occurs before a

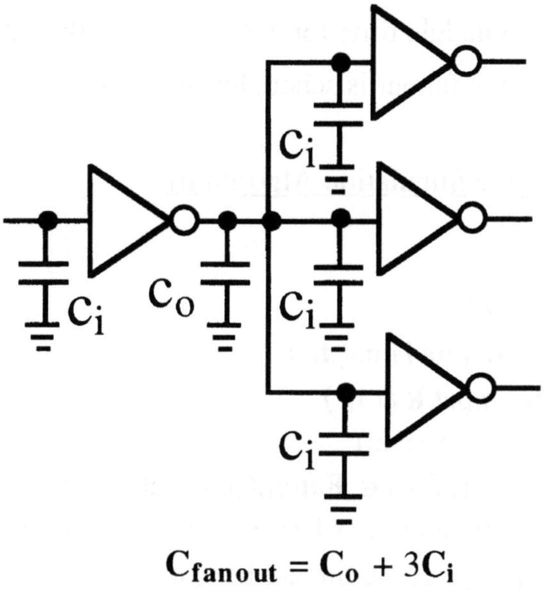

Figure 5.16: Computing the Total Node Capacitance

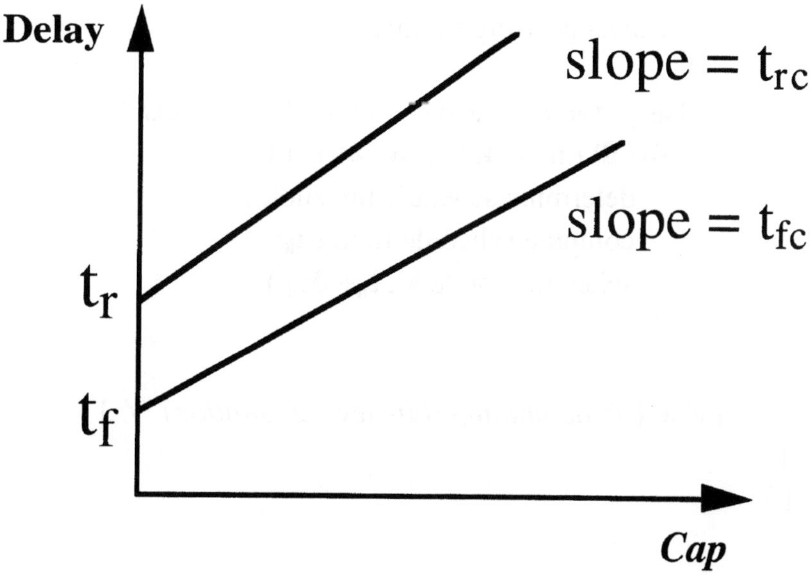

Figure 5.17: Delay vs. Capacitance for an Inverter

transition in the opposite direction is completed, a glitch warning is produced and the original transition event is cancelled. If the event does not cause a glitch, the schedule time for each fanout of the output node, $\mathbf{\Delta t_j}$, is computed and then the fanout is scheduled at $\mathbf{t_n} + \mathbf{\Delta t}$.

<u>**Algorithm 5.1** (Logic Simulation Algorithm)</u>

```
tn← 0;
while (tn≤Tstop) {
    tn← NextEventTime( tn );
    foreach ( input k at tn )
        if ( ek is active )
            forall ( j ∈ Fanout(ei) ) schedule( node j, tn );
    foreach ( node i at tn ) { /* processing logic block i */
      get input_states; compute new_output;
      if (node i has changed ) {
        compute delay, Δt;
        if ( current_time < last_event_time(i)) {
          issue glitch message;
          cancel pending events;
        }
        else { /* normal event, so schedule fanouts */
          forall ( node k ∈ Fanout(i) ) {
            determine schedule threshold;
            compute schedule time Δtk;
            schedule ( node k , tn+ Δtk );
          }
        }
      } else { /* do nothing (latency exploitation) */ }
  }
  □
```

CHAPTER 6

SWITCH-LEVEL TIMING SIMULATION

6.1. INTRODUCTION

Most modern logic simulators handle the problems specific to MOS integrated circuits by including the notion of signal strength in the logic model. However, the use of strength does not, by itself, solve all the modeling problems inherent to MOS circuits. For example, circuit designers use many combinations of transistors which do not have a direct mapping to a logic gate and therefore cannot be represented conveniently at the gate level. It is also difficult to model the logic operation of dynamic circuits in a convenient form in a standard logic simulator. Transfer gates further complicate the situation because they introduce dynamic loading effects, bidirectional signal flow, and capacitive charge-sharing effects. Many of these problems were resolved with the advent of the switch-level modeling and simulation technique [BRY80].

This chapter begins with a description of standard switch-level simulation and identifies a number of limitations in the approach, primarily the lack of accurate timing information, and also the fact that intermediate voltage states are not represented which may occasionally lead to incorrect results. Then, an electrically-oriented switch-level modeling technique that resolves these and other problems is described. This technique allows variable precision simulation, thereby allowing the user to choose anywhere from logic simulation accuracy to electrical simulation accuracy. Hence, the approach effectively spans the "gap" between logic and electrical simulations. A number of other simulation approaches with similar properties are also described. In the last section, the use of this variable precision modeling approach to map signals across the interface between logic gates and electrical circuitry is described.

6.2. SWITCH-LEVEL SIMULATION

A switch-level simulator transforms an MOS transistor network into a corresponding network of switches and performs logic simulation on the resulting network. For example, in MOSSIM [BRY80], the logic circuit is described entirely at the transistor level, and the transistors are modeled as simple gate-controlled switches. The switch-level logic state model includes three logic levels (0, **X**, 1) and a number of strengths, **s**, which lie in the range $\{1, \cdots, w\}$. Two subranges of strengths are defined, one representing all signal strengths originating at some external source, in the range $\mathbf{k} < \mathbf{s} < \mathbf{w}$, and the other corresponding to nodal capacitance values, in the range $1 \leq \mathbf{s} \leq \mathbf{k}$. The maximum possible strength, **w**, is reserved for inputs only. The switch-level model attempts to incorporate the key aspects of MOS logic circuits that determine its behavior and abstract away the details of the electrical behavior. This approach greatly simplifies the algorithms needed to correctly simulate a large variety of MOS logic circuits.

The simulation process in switch-level simulation proceeds as follows. First, as a preprocessing operation, the switch-level network is partitioned into a number of subnetworks which are collections of *strongly-connected components* (SCC) or *channel-connected components*. These are sets of transistors that are connected to one another at the source or drain terminals. The identification of SCCs can also be done dynamically during the simulation process. Processing a given SCC may require a complicated series of steps, possibly involving iterations, to account for the interactions of different strengths of two or more "ON" transistors, as described below. However, the interaction between two SCCs is easier to analyze since they are connected at the gate inputs of transistors and, hence, the logic operations do not depend on the signal strengths. The SCCs are simply scheduled and processed in the manner described earlier for logic gates using event-driven, selective trace techniques. Therefore, this mode of simulation well-suited to implementation in mixed-mode simulators.

The complicating factor in the processing of SCCs is due to the bidirectionality of transfer gates, or pass transistors. Although the transfer gate is inherently a bidirectional element, it is usually found in applications in which the signal flow is intended to be unidirectional. That is, the circuit designer expects signals to flow in only one direction through the device. However, there are occasions when transfer gates are used in bidirectional applications, or other situations in which a design error leads to signal flow in different directions at different times. A simulator must be able to analyze these cases accurately if it is to be useful. There have been a variety of modeling approaches for bidirectional transfer gates, including the unconventional approach of two unidirectional elements connected back-to-back. This approach can lead to inconsistencies when different logic values are on opposite sides of the element. Each value can flow through one of the transfer gates and reach the opposite side and then propagate through the circuit producing incorrect results.

During the evaluation of an internal node of an SCC, the elements connected to that node try to impose their values on the node and the final state is determined by the element with the highest strength. A path analysis is actually performed to identify all possible paths from the node to a supply or ground node [BRY84]. Each path is assigned a strength that depends on the transistor with the lowest strength. Weak paths are blocked at intermediate nodes if a stronger path is encountered at the nodes. The strongest path to a given node determines the final state of the node. If two paths of equal strength but opposite values are encountered at a given node, the node is assigned to the **X** state.

The path analysis approach is effective for handling bidirectional signal flow. A simpler approach is to use the so-called "supernode" technique [BRY80]. In this approach, all nodes that are connected through transistors that are "ON" are considered to be the same node for processing purposes. All devices connected to this composite node are processed together to determine the new state. The new state is then assigned to all nodes which comprise the supernode. The main problem with this approach is that it cannot adequately handle the case where the final values at the nodes are different and determined by the strengths of the transistors in the subnetwork. Therefore, it does not permit different nodes of a supernode to reach different values.

An alternative to the supernode and path analysis approaches is to use an iterative or relaxation-based method to determine the new states of these strongly-connected nodes (SCN) [DUM86]. The first step in this approach is to assign all nodes to the lowest strength permissible, or to a strength associated with the capacitance at each node. The signals are then propagated from the source nodes through the switch network starting from the signal possessing the largest strength. This processing order prevents the accidental propagation of weaker signals onto storage nodes that may inadvertently generate the unknown logic level. The internal nodes of the SCC are evaluated using local event-driven techniques and the fanout nodes within the set of SCNs are scheduled whenever they change state. The process is repeated until convergence is obtained, at which point scheduling occurs at the SCC level, i.e., the fanout SCCs are scheduled. Note that by using iterative methods, the nodes within a strongly connected component may converge to different logic levels, which is the main advantage of the approach. The combination of local relaxation methods at the SCN level and standard event-driven methods at the SCC level allows efficient switch-level analysis to be performed.

One problem not addressed above is that of processing transfer gates with unknowns at gate inputs (i.e., **X**-transistors). The strategy in analyzing circuits with **X**-transistors is to minimize the number of **X** states generated at the internal nodes of an SCC. A pessimistic approach would be to generate the **X** state at each of the drain and source output nodes of each **X**-transistor. This method is the easiest to implement but it may actually force the simulator to process many more events than necessary since the **X** state tends to be "sticky" and propagates throughout the circuit very quickly [CHA87]. A brute-force approach would be to enumerate all possible combinations of gate input values by replacing the **X**-transistors by either 1-transistors or 0-transistors. If there are **k** **X**-transistors, the SCC would have to be evaluated a total of $2^{\mathbf{k}}$ times! Any node which produces the same logic level, regardless of the input combination, is set to that logic level; otherwise, it is set to **X**.

A better approach [BRY87], which offers linear computational complexity in **k**, is to first choose the gate settings of the **X**-transistors to maximize the number of 1's or **X**'s in the SCC under consideration. Then, the process is repeated to select the gate settings to maximize the number of 0's or **X**'s in the SCC. Again, any node which reaches the same logic level in both cases is set to that level; otherwise, it is set to **X**. This approach has been shown to produce the same results as the computationally expensive method described above but requires much less work.

6.3. A GENERALIZATION OF THE NINE-STATE LOGIC MODEL

Switch-level simulation has been adopted as an efficient technique for functional verification of large MOS digital circuits. However, there are many circuit configurations that may lead to incorrect solutions when the simple switch-level model is used. In fact, simple examples can be constructed that require more than the three strengths and three states of the nine-state logic model (described in Chapter 5) to produce the correct solutions. To illustrate this point further, consider the two circuit fragments shown in Fig. 6.1. The circuit in Fig. 6.1(a) is a 2-ϕ regenerative latch driving a bus. The two inverter stages provide forward gain while

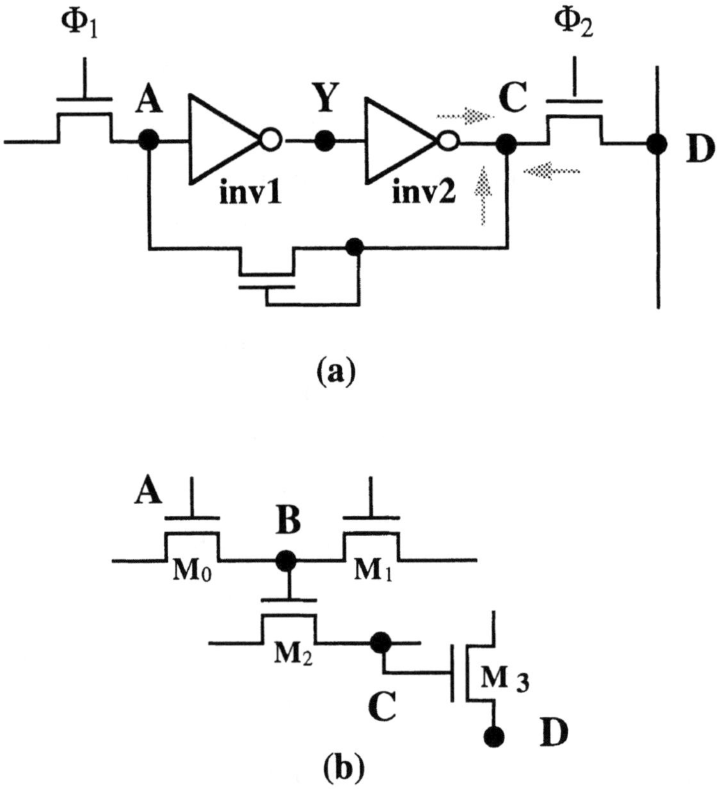

Figure 6.1: Examples Requiring Additional States and Strengths

the depletion load device provides a resistive feedback path from **C** to **A**. In this case, at least four strengths are necessary to obtain the correct results at node **C**: inverter inv2 has a weak pull-up strength $\mathbf{W}_1$, a forcing pull-down strength **F** and a high-impedance pull-down strength **H** (depending on whether it is on or off). The depletion device has a resistive strength $\mathbf{W}_3$, such that $\mathbf{W}_3$ is less than $\mathbf{W}_1$. In addition, the pass transistor connected to the bus also has a strength, $\mathbf{W}_2$, which is less than $\mathbf{W}_1$ but greater than $\mathbf{W}_3$. As described earlier, switch-level simulators provide a range of strengths to address this problem.

A more serious problem is that the switch-level model may not produce the correct results for an arbitrary connection of pass transistors when threshold voltage drops are important. For example, the situation shown in Fig. 6.1(b) is a case where additional voltage states are necessary. Here, the designer has inadvertently connected the gate of transistor $\mathbf{M}_3$ to a node which is already two threshold voltage drops below the input signal. Therefore, the value at node **D** can only rise to three threshold voltage drops below the input value and this may not be high enough to be considered as a valid high. While this circuit is clearly a poor design, it is important for a simulator to detect this type of error. A standard switch-level simulator would not be capable of identifying this error since it uses only three logic values. For this case, at least three additional logic values are required. Hence, an appropriate state model to adequately simulate both circuits in Fig. 6.1 is shown in Fig. 6.2.

Another situation which requires multiple strengths and voltage levels is the simulation of dynamic circuits. In these circuits, capacitive charge-sharing and feedthrough effects often degrade the voltage levels. Feedthrough usually occurs when a clock signal feeds through a floating capacitor to an isolated node with a grounded capacitance. The isolated node also sustains a sharp voltage transition which depends on the value of the grounded and floating capacitance values. Usually, feedthrough is not a significant factor. However, charge-sharing can often lead to circuits that do not function properly. Charge-sharing occurs when a transistor connecting two isolated grounded capacitors is turned on. The total charge is redistributed between the two capacitors until their node voltages are equal. It is possible to handle charge-sharing without introducing additional logic levels by assigning to each node a strength that corresponds to its capacitance value. If charge-sharing occurs, the node with the larger capacitance imposes its value on the node with the smaller capacitance (for worst-case analysis) and a potential problem is at least observed [BRY80].

One basic limitation of standard switch-level simulation still

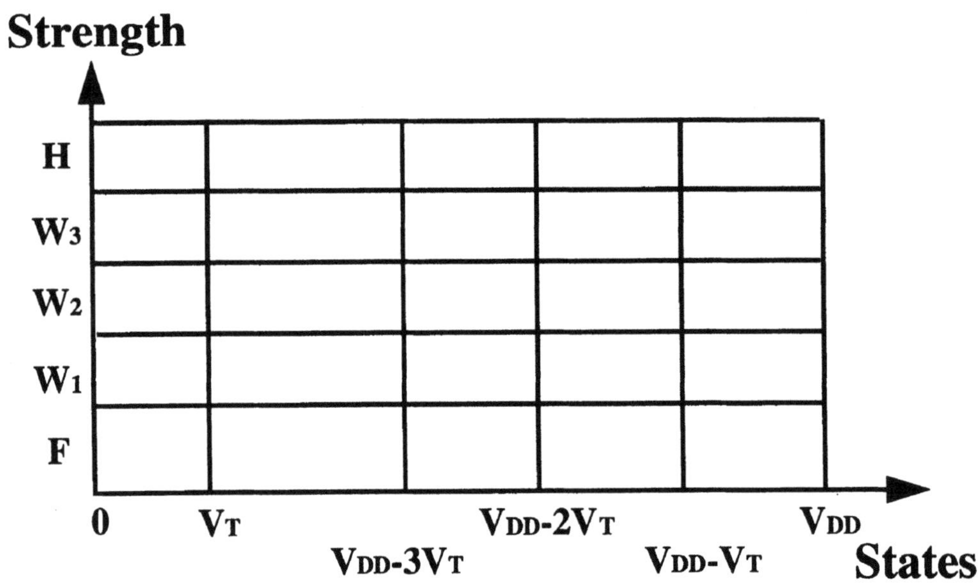

Figure 6.2: Better Logic Model for Simulation of Circuits in Fig. 6.1

remains: accurate timing information is not provided. Electrical simulation provides detailed timing information but is very expensive due to the use of complex analytical models that characterize the transistor current-voltage relationships. Logic simulation is extremely fast but is often unable to provide more than first-order timing information using simple expressions to compute the rise and fall delays. Clearly there exists a large "gap" between electrical simulation and logic simulation. The arguments made above promoting multiple logic values and strengths and the requirement for switch-level timing simulation can be resolved by treating strengths as electrical resistances and logic states as electrical voltage levels. This connection allows a generalization of the model of strength vs. state in logic simulation to resistance, **R**, vs. voltage, **V**, in electrical simulation.

The **R–V** characteristics for an inverter driving a pass transistor are shown in Fig. 6.3 based on SPICE2 simulations. These are dc transfer curves of the output resistance of the inverter and the output resistance of the transfer gate as a function of their respective output voltages. The two curves are highly nonlinear and do not exhibit monotonic behavior. Conceptually, electrical simulation has an infinite number of allowed "states" in this plane while the higher levels of simulation discretize the horizontal and vertical axes into a finite number of states. As a result, the difference between electrical and logic simulations becomes one of the degree of discretization of the **R–V** plane. This relationship also provides a convenient way of mapping from one form of simulation to the other in mixed-mode simulation. The use of this model as a vehicle for simulation is described in the next section. Its application in signal mapping across the mixed-mode interface is addressed in Chapter 7.

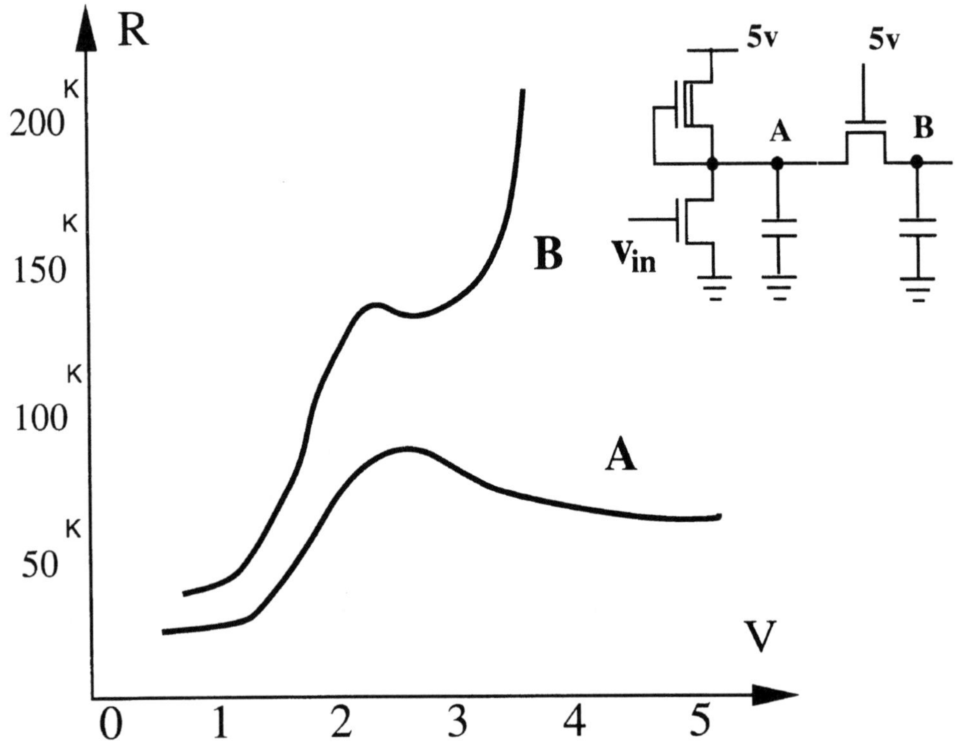

Figure 6.3: Resistance vs. Voltage Plane

6.4. SIMULATION USING THE GENERALIZED MODEL

6.4.1. Electrical-Logic Simulation

A variable precision simulation approach, called *electrical-logic simulation* or simply *Elogic* [KIM84], has been developed based on the generalized model described in the previous section. This form of simulation can be viewed as a relaxation-based, switch-level simulation technique. Elogic uses electrical device models in the context of switch-level simulation which allows electrical timing information to be obtained. As part of the Elogic modeling process, a number of discrete voltage levels are selected. These levels need not be equally spaced but the number of levels and their values have an impact on performance and accuracy. In standard electrical simulators, the time-step is selected first and then the node voltage change is computed. By contrast, in Elogic the voltage step is known in advance and the *time* required to make a transition from one voltage state to another adjacent voltage state is computed. Similar approaches are used in SPECS [DEG84], MOTIS3 [TSA85], SPECS2 [VIS86] and ADEPT [ODR86], as described in the next section.

The processing sequence in Elogic is illustrated in Fig. 6.4 for a simple inverter example. The set of Elogic states is defined to be $\mathbf{V}_0$, $\mathbf{V}_1$, $\mathbf{V}_2$, $\mathbf{V}_3$, and $\mathbf{V}_4$. As shown in Fig. 6.4(b), the input makes a sequence of transitions from $\mathbf{V}_0$ to $\mathbf{V}_4$ and visits each intermediate state between the two end points. Each transition at the input node causes an event to be scheduled at the output node. The corresponding output computed by Elogic is illustrated in Fig. 6.4(c). Note that the first transition at the input does not cause a transition at the output node since the transistor does not turn on. However, the second transition and all subsequent input transitions result in transitions at the output. Note also that the output continues to make transitions even after the input reaches its final value due to a self-scheduling mechanism.

Briefly, the steps required to calculate the transition time, $\mathbf{\Delta t}$, are as follows: each nonlinear device is first replaced by a linearized equivalent model. This model is used to compute the steady-state or final voltage, $\mathbf{V}_{ss}$. An exponential characteristic is used to predict the transient behavior of the voltage at the output node from the present state, $\mathbf{V}_n$, to the final value, $\mathbf{V}_{ss}$. The transition time, $\mathbf{\Delta t}$, is then computed as the time required to go from $\mathbf{V}_n$ to $\mathbf{V}_{n+1}$ along this exponential characteristic. After the input has completed its sequence of transition events, the output still continues to be scheduled due to its own self-scheduling mechanism, similar

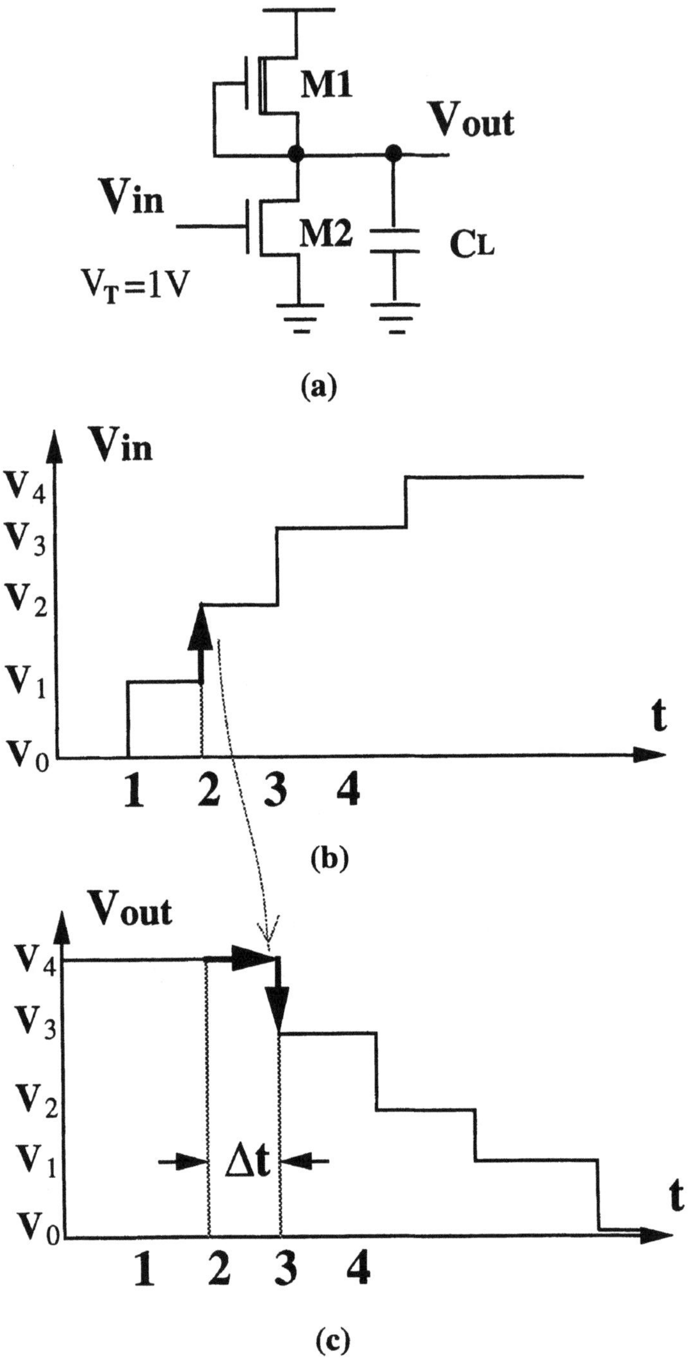

Figure 6.4: Elogic Processing Sequence for a Simple Inverter

to the one described for ITA in Chapter 4. The output will continue to schedule itself until it reaches the steady-state level.

The algorithm is modified slightly if the input makes a new transition *before* the output has completed its current transition. This situation is usually categorized as a glitch in logic simulation but it calls for the rescheduling of a pending event in Elogic. If the output is very close to the next state, $\mathbf{V}_{n+1}$, it is set to the next state and a new event is scheduled only if a transition to $\mathbf{V}_{n+2}$ is warranted. If the output is still very close to the previous state, $\mathbf{V}_n$, it is reset to the previous state and a new event time is calculated for the transition to $\mathbf{V}_{n+1}$. If the output is somewhere in between the two states, a new transition time is calculated using $\mathbf{V}_n$ and the new value of the input node. The event is then rescheduled at the average of the original event time and the new event time.

The number of Elogic voltage levels selected and their position have an important impact on the accuracy and speed of simulation. Specifically, the precision with which a given voltage can be represented is limited by the set of voltage levels chosen in an Elogic model. If the actual value of a node voltage is between two Elogic states, the node voltage must be set to the closest defined level. This operation is analogous to a roundoff process and it produces a roundoff error. The number of states can always be increased to improve the precision in representing a particular voltage. However, since it is necessary to visit each intermediate state whenever a transition is made from some initial state to the final state, the simulation time increases as the number of states increases. It is this continuous tradeoff between speed and precision that makes Elogic particularly attractive as it effectively spans the large speed/precision "gap" between classical electrical and logic simulations. The user can use very few states in the preliminary design phase to verify the functionality of the circuit and obtain crude first-order timing estimates. As the design matures, more and more states can be added as necessary to improve the accuracy of the analysis. In addition, different parts of the same circuit can be simulated using a different number of states; this constitutes mixed-precision simulation, which is a special form of mixed-mode simulation.

The detailed calculations for the transition time are now described using the two nonlinear devices connected to a linear grounded capacitor given in Fig. 6.5(a). Assume that the initial state of the node is $\mathbf{V_n}$. When the output node is processed, the nonlinear devices are first converted to linear devices. This can be done using either a small-signal model, which uses the incremental conductance and current of the device relative to a given operating point, or a line-through-origin model which uses the large-signal conductance of the device. In either case, the model is obtained by a table lookup scheme. The linear equivalent network following this step is shown in Fig. 6.5(b). From this circuit, it is clear that the steady-state value of the output node is

$$\mathbf{V_{ss} = I_N/G_N} \tag{6.1}$$

where $\mathbf{I_N} = \sum_{i=1}^{n} \mathbf{I_i}$ and $\mathbf{G_N} = \sum_{i=1}^{n} \mathbf{G_i}$. The next step is to determine if the node will undergo a transition from the present state to another adjacent state. This is done by checking if either:

$$\mathbf{V_{ss} > V_n} \tag{6.2a}$$

or

$$\mathbf{V_{ss} < V_n} \tag{6.2b}$$

If either condition is true, a transition time calculation is warranted.

The dynamic behavior of the linear equivalent circuit is given by

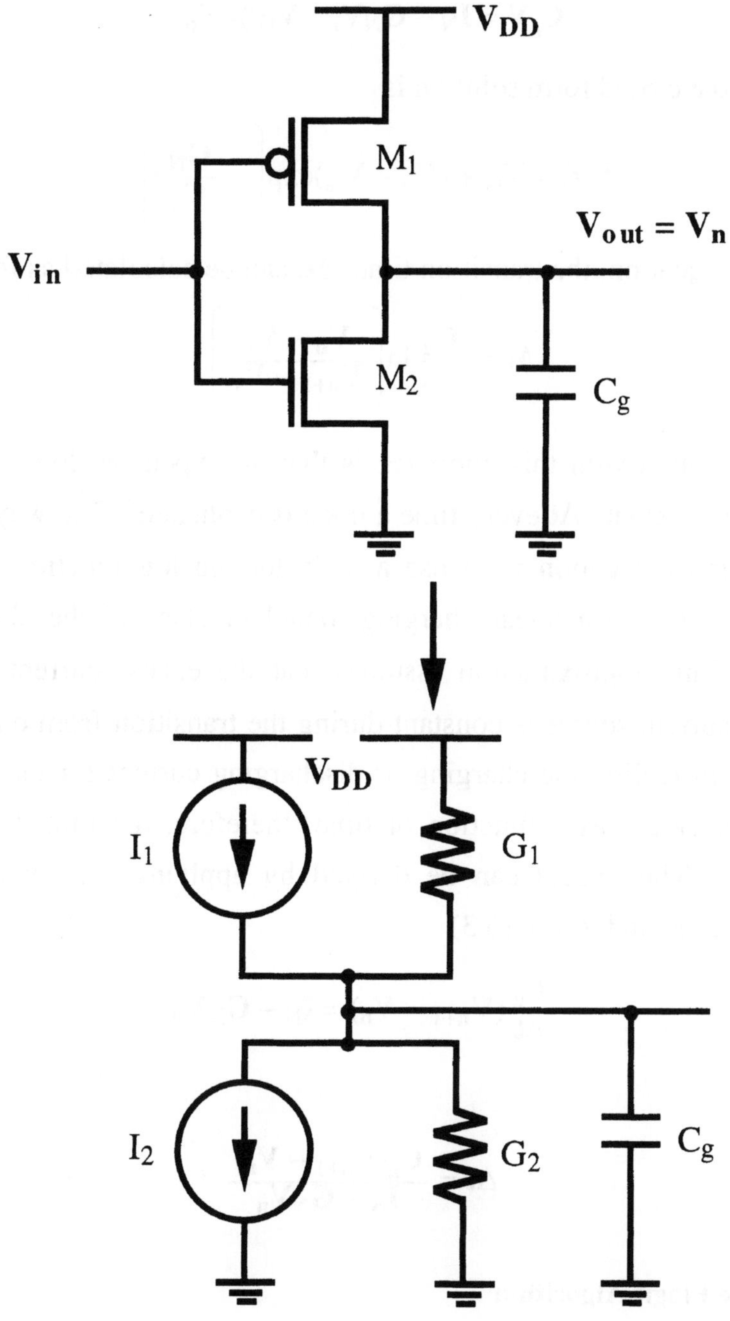

Figure 6.5: Circuits Used to Calculate Transition Time

$$C_g\dot{V} = I_N - G_N V, \quad V(0)=V_n \tag{6.3}$$

for which the closed form solution is

$$V(t) = V_{ss} + (V_n - V_{ss})\exp\left[-\frac{G_N}{C_g}t\right]. \tag{6.4}$$

Using this equation, the transition time, Δt, can be calculated as follows:

$$\Delta t = \frac{C_g}{G_N}\ln\left[\frac{V_n - V_{ss}}{V_{n+1} - V_{ss}}\right] \tag{6.5}$$

A problem with this approach is that an expensive log function is required to calculate Δt every time a node is evaluated. One way to avoid this function evaluation is to use a table lookup log function. Another approach is to use a linear charging model in place of the closed-form solution. This approximation assumes that the excess current available from the current source is constant during the transition from one state to the next. In reality, the charging or discharging current for the capacitor tends to decrease as a function of time; therefore, the model is always optimistic. This model can be derived by applying the forward-Euler integration method to Eq. (6.3)

$$\frac{C_g}{\Delta t}(V_{n+1} - V_n) = I_N - G_N V_n \tag{6.6}$$

Then,

$$\Delta t = \frac{C_g(V_{n+1} - V_n)}{I_N - G_N V_n} \tag{6.7}$$

6.4.2. The Elogic Algorithm

The details of the Elogic simulation algorithm are presented below. In the algorithm, Elogic is implemented using event-driven techniques since a node schedules its fanouts for processing only when it achieves a new state. If a fanout node has already been scheduled at a some time $\mathbf{t_E}$, in the future, it will be rescheduled at the present time, $\mathbf{t_i}$.

Algorithm 6.1 (Electrical-Logic Simulation Algorithm)

```
/* processing node i */
  if ( ti=tn+1 OR Vi≈Vn+1 ) { /* reached new state */
      recompute ← FALSE;
      update voltage, Vi←Vn+1;
      /* fanout scheduling */
      forall ( fanout nodes k of node i )
         schedule ( node k at time ti );
  }
  else { /* did not reach new state */
      reset voltage, Vi←Vn;
      reset time, ti←tn;
      if ( Vi≈Vn ) recompute ← FALSE;
      else recompute ← TRUE;
  }
  GN←0 and IN←0;
  forall (fanin nodes k of node i) {
      replace node k by a constant voltage source;
      compute Gk and Ik;
      update GN and IN;
  }
  compute Vss, the steady-state voltage;
  /* Check for transition using Eq. (6.2) */
  if ( node i can make a transition ) {
      transition ← TRUE;
      compute transition time, Δt, using Eq. (6.2) or (6.7);
      if (recompute = TRUE) Δt=(Δtn+Δtn+1)/2;
  }
  else transition ← FALSE;
  if ( transition = TRUE ) {
```

```
            if ( (t_i + Δt) < T_STOP ) /* self-scheduling */
                schedule ( node i at t_i + Δt );
        }
        else {do nothing}; /* latency exploitation */
```

□

6.4.3. Problems with the Elogic Approach

The Elogic algorithm, if implemented exactly as described above, may encounter certain problems that lead to excessive computer run times or reduced accuracy. The first problem is that of algorithmic oscillation of a node voltage where one does not exist in the true solution. The simple form of this problem arises if the steady-state solution, $\mathbf{V}_{ss}$, lies between two discrete Elogic states. For example, if $\mathbf{V}_{ss}$ lies in the range $\mathbf{V}_1 < \mathbf{V}_{ss} < \mathbf{V}_2$, then the node will be assigned to the value $\mathbf{V}_1$ or $\mathbf{V}_2$, whichever is closer to $\mathbf{V}_{ss}$. However, if the node is re-evaluated using the new Elogic state voltage, it may force the node to move in the opposite direction, in which case it will be set to the other neighboring state. Again, since the true solution is in between the two defined states, the node will attempt to make another transition in the opposite direction creating an oscillation situation.

One approach to resolve this problem is to detect oscillations and then suppress them. This may lead to the inadvertent suppression of actual oscillations in the circuit; therefore, it is not an attractive solution. Another approach is to introduce hysteresis into the state transition criterion whenever the node voltage changes direction. Simple oscillation usually occurs as the steady-state voltage is reached. Therefore, if the sign changes on the time derivative of voltage, it is appropriate to require a significant change in the value before a transition in the opposite direction is undertaken. For example, the transition could be scheduled if the new steady-state voltage is beyond the midpoint of voltage region just visited during the last transition. Using this strategy, a transition occurs only if $\mathbf{V_{ss}} > (\mathbf{V_n}+\mathbf{V_{n+1}})/2$ or $\mathbf{V_{ss}} < (\mathbf{V_n}+\mathbf{V_{n-1}})/2$. Another way to resolve the problem is to simply set the node to an intermediate "illegal" voltage level when the steady-state interval is reached. The node is permitted to leave this illegal state only if it is scheduled by another node.

There is a second source of oscillation, termed *interactive oscillation*, which is more insidious and involves two or more nodes. As shown for the circuit in Fig. 6.6(a), the problem occurs when two neighboring nodes use each other's values to determine their next states and the true solution lies between two Elogic states. In this case, node A is scheduled to make a transition from 1V to 0V, while node B is scheduled to make a transition from 0V to 1V. However, after the transitions occur, both nodes make a transition in the opposite direction, and this process continues indefinitely. This type of oscillation is more difficult to detect than the simple oscillations described earlier, but the problem can be solved by introducing more states into the Elogic model.

A third problem arises due to strong coupling between two or more nodes in the circuit. This problem can be illustrated using a simple circuit as shown in Fig. 6.6(c), where $\mathbf{G_1}$=**1mho** and $\mathbf{G_2}$=**9mhos**, and initially $\mathbf{V_A} = \mathbf{V_B} = 0$ V. Note that in evaluating node **A**, a zero volt source is applied at node **B** thereby grounding it. The Norton equivalent model seen by node **A** is computed as follows:

$$\mathbf{I_N} = 5\ \mathbf{G_1} = 5 \times 1 = 5\ (\mathbf{A}) \tag{6.8a}$$

$$\mathbf{G_N} = \mathbf{G_1} + \mathbf{G_2} = 1 + 9 = 10\ (\mathbf{mhos}) \tag{6.8b}$$

Therefore, the Thevenin equivalent voltage is

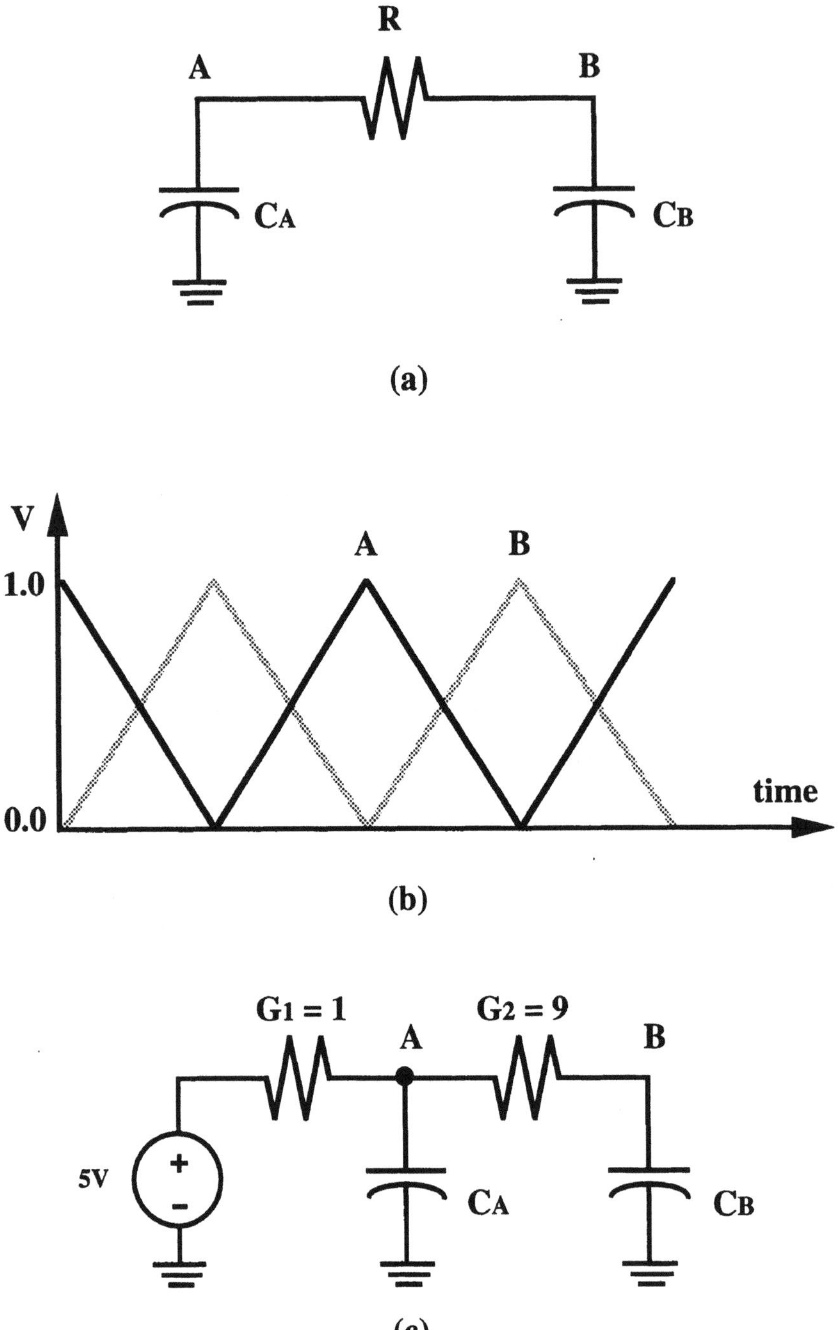

Figure 6.6: Simple Elogic Problems
(a) example circuit (b) interactive oscillation (c) strong coupling

$$V_{ss} = I_N / G_N = 0.5 \text{ V} \tag{6.8c}$$

Clearly, if the voltage change necessary to warrant a transition is larger than 0.5V, the basic Elogic method would not attempt to transfer node A to the next adjacent state. As a result, both V_A and V_B would remain at zero volts. As described earlier, strong coupling affects the convergence speed of ITA and WR, whereas in the case of Elogic, it results in a transition error. For this circuit, the maximum voltage step which can be used depends on the ratio of G_1 and G_2 and, in general, the Elogic states for a given problem should be selected with this rule in mind. Another solution to this problem is to determine the steady-state voltages of all nodes in a set of SCNs using switch-level techniques, and then schedule transitions based on this analysis [TSA85].

6.5. A SURVEY OF SWITCH-LEVEL TIMING SIMULATORS

A number of other switch-level timing simulation techniques have been developed over the past decade that are also appropriate for use in a mixed-mode simulator. The original work in this area was, of course, the timing simulation algorithms of MOTIS [CHA75] as described in Chapter 4. More recently, there have been a number of notable contributions that are embodied in the programs RSIM [TER83], SPECS [DEG84], MOTIS3 [TSA85], ADEPT[ODR86], SPECS2 [VIS86], and iDSIM [RAO89]. The techniques used in these programs are reviewed briefly below.

The RSIM program attempted to produce timing waveforms for the switch-level technique by adding a linear resistor in series with each transistor switch and providing a capacitor to ground at each node. The value of the resistor was set to infinity when the gate voltage was low and to some finite resistance when the gate was high. Resistance values were calculated using the length and width of the transistors. The logic state model included only 0, **X** and 1. During the simulation, the transistors were replaced by their equivalent resistances and then combined to form a Thevenin equivalent circuit, with resistance $\mathbf{R_{drive}}$ and voltage source $\mathbf{V_{thev}}$, driving a loading capacitance, $\mathbf{C_{load}}$. When a transition was expected at a node, the time required to make the transition was computed as $\mathbf{R_{drive} \times C_{load}}$. Since the values of $\mathbf{R_{drive}}$ for low-to-high transitions and high-to-low transitions were computed using different values of resistances for the transistor, the accuracy was often within 30% of SPICE2 for many circuits while providing over two orders of magnitude of speed improvement. The Elogic method can be viewed as an extension of this approach with the flexibility of allowing more states and having table lookup equivalent models for the transistors in each state.

The "fast timing" simulation approach of MOTIS3 is based on the Elogic algorithm. However, a backward correction scheme is used with the variable voltage step scheme to improve accuracy and avoid oscillatory behavior. First, the net current, $\mathbf{I_{net}}$, available to charge the load capacitance, $\mathbf{C_{load}}$, is calculated. Then the time required to make the transition is calculated using either an exponential model or the forward-Euler model (shown here):

$$\mathbf{h_n = (V_{n+1} - V_n)/I_{net}} \qquad (6.9)$$

Next, the value $\mathbf{h_n}$ is used to perform a regular integration step to compute a new target voltage, $\mathbf{V'_{n+1}}$. Finally, $\mathbf{h_n}$ is scaled to produce the actual event time:

$$\mathbf{h'_n = h_n(V'_{n+1} - V_n)/(V_{n+1} - V_n)} \qquad (6.10)$$

One additional contribution in MOTIS3 is the use of a so-called "superblock" approach to handle tightly-coupled nodes. First, the steady-state voltage, $\mathbf{V_{ss}}$, of every node in the superblock is computed. Then, the delay is calculated for each node and the minimum delay is assigned to the superblock. The node voltages in the superblock are scaled with respect to this delay.

The ADEPT approach is also similar in many ways to the basic Elogic approach described in the previous section. Like the MOTIS3 approach, it allows variable voltage steps to be used to improve accuracy at the expense of additional CPU-time. However, its most distinguishing feature is the implicit dynamic partitioning approach used to process tightly-coupled nodes. In ADEPT, when a node **i** is computed, the nodes, {**j**}, that are neighbors of **i** are checked for tight coupling to **i** using the criterion:

$$\frac{\mathbf{C_{ij}\dot{V}_j + G_{ij}V_j}}{\mathbf{I_i}} > \varepsilon \qquad (6.11)$$

All neighboring nodes that satisfy this criterion are solved using local relaxation methods to produce the correct results. Since this is applied to every node separately, it can be viewed as overlapped, dynamic partitioning.

Another promising variable precision approach has been implemented in the SPECS2 program, which is based in part on the techniques used in SPECS [DEG84]. A tree/link based equation formulation [CHU75] is used in the program, instead of the standard nodal formulation described in Chapter 2. This approach is well-suited to the simulation of circuits containing ideal switches that have infinite resistance when OFF and infinite conductance when ON. Devices with these properties are very troublesome in the context of nodal analysis. In tree/link based analysis, a circuit graph is constructed from the circuit description and a tree is identified in the graph. A *tree* is defined as a connected, acyclic subgraph that contains all the nodes of the original graph. The branches that belong to the graph are called *tree branches* while the remaining branches are the *links*. The links combine to form a *cotree*. Once the tree has been defined, a *cutset* is identified in the tree. A cutset is some subset of the branches of a tree such that their removal results in a graph that is no longer connected, but the insertion of any one of the branches from the removed set results in a connected graph. Cutsets are the subgraphs to which KCL is applied, and loops are the subgraphs to which KVL is applied. The fundamental cutsets and loops are used to formulate the circuit equations.

SPECS2 uses table models to define the device I-V characteristics as shown in Fig. 6.7(a). Note that the segments are piecewise constant, forming a set of step functions, as opposed to being piecewise linear. These steps are important in the event-driven approach of SPECS2 since an event occurs whenever a device reaches a "corner" of its step model, as shown in Figs. 6.7(b) and 6.7(c). Here, events occur at $\mathbf{t}_1$, $\mathbf{t}_2$ and $\mathbf{t}_3$ since $\mathbf{V}_1$, $\mathbf{V}_2$ and $\mathbf{V}_3$ are all boundaries of the device step model. For example, at $\mathbf{t}_2$ there is a change in the current through the device from $\mathbf{i}_1$ to $\mathbf{i}_2$. As a result, the corresponding device is processed with the new current value and the next event is scheduled at the next corner in the table. The effect of this change on the rest of the circuit is taken into account via subsequent event scheduling and processing.

The SPECS approach is also prone to spurious oscillations, as are many other variable precision algorithms. The strategy used in SPECS2 to overcome this problem is to place the element into a pseudo-steady- state condition. This is done by picking a current for the device which is *in between* the currents in the table model whenever the direction of the current derivative (with respect to time) changes sign. The value is selected to place the device in steady state. If the device is truly in steady state, it will remain in this condition. On the other hand, if it is not, it will be forced out of it by the other elements in the circuit. Therefore, true circuit oscillation will not be suppressed but algorithmic oscillation will be

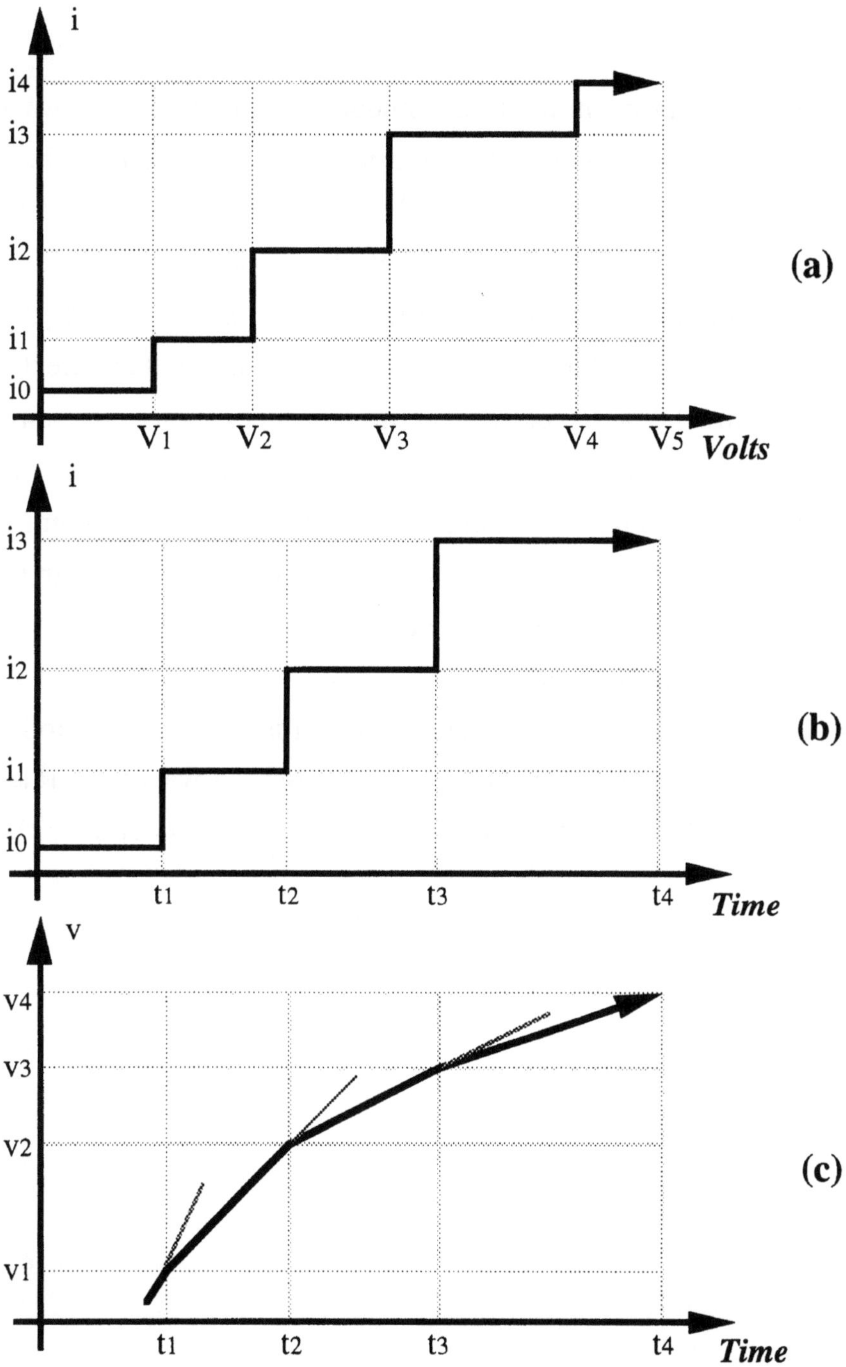

Figure 6.7: Table Models and Simulation Events in SPECS2

prevented.

The iDSIM program uses macromodeling combined with waveform relaxation to perform switch-level timing simulation. The switch-level network is preprocessed to identify series-parallel connections of transistors to form composite transistors. A set of macromodel parameter tables is generated for each composite transistor based on the device gate voltage, device size, threshold voltage, and other factors that affect delay. The actual simulation is performed in two steps. First the circuit is analyzed using switch-level techniques to identify the transitions that will occur during the simulation. These transitions produce break points in the waveforms that, in turn, define the time intervals for detailed simulation. The second step is to perform the delay analysis to compute the transition times using the tables generated for the devices. If there are no feedback loops in the circuit, one pass of this algorithm is sufficient to produce the desired results. When feedback loops are present, a waveform relaxation approach is used with partial waveform convergence to compute the final results.

CHAPTER 7

THE MIXED-MODE INTERFACE

A major issue in all mixed-mode simulators is the problem of interfacing of two or more simulation modes. This chapter describes the modeling problems of these interfaces and possible solutions. Of course, this problem arises only when elements from different modes of simulation are connected at a common node. There are two possible directions of signal conversion: one from the lower level of simulation (more detailed) to the higher level of simulation (less detailed), and a second going in the opposite direction. For example, logic simulation and electrical simulation require signal conversions from logic to electrical simulation and from electrical to logic simulation. Typically, it is easier to translate a signal from a lower level of simulation to a higher level since the conversion involves removing unnecessary details from the signal. Signal conversions from a higher to a lower level are more difficult since the information required at the lower level is often unavailable at the higher level. In particular, it is the conversion of signals from the logic domain to the electrical domain that is most troublesome. Both types of signal conversions are described in this chapter. A number of examples are used to illustrate potential problems of the various interface models in use today.

7.1. ANALOG TO DIGITAL INTERFACE

In a typical analog to digital conversion process, the analog voltage waveforms are transformed into equivalent digital waveforms using *thresholding functions*. This operation discretizes a continuous function based on a set of predefined voltage thresholds. Specifically, two threshold voltages, $\mathbf{V_{IH}}$ and $\mathbf{V_{IL}}$, are chosen according to the dc voltage transfer characteristics (VTC) of the logic gate, as described in Chapter 5. Then, if the input to a logic gate is a rising (falling) waveform that reaches $\mathbf{V_{IL}}$ ($\mathbf{V_{IH}}$), the output waveform of the logic gate begins to fall (rise) to a logic 0(1) at that point in time. This type of signal conversion works reasonably well so long as the analog input waveform is well-behaved in terms of driving a logic gate.

There are, however, two cases that cause problems at this interface due to the behavior of the analog waveform. First, if the analog input waveform settles at a voltage between $\mathbf{V_{IL}}$ and $\mathbf{V_{IH}}$, the correct logic output becomes an issue. According to the algorithms used in traditional logic simulators, an unknown logic value, or **X** state, should be reported to the user. If the VTC-based switching described above is used, the output undergoes a full logic transition even though the input settles at some midrange value. In either case, the result may be misleading to the designer. Another problem arises if the input waveform does not have monotonic behavior in time, in which case the output of the logic gate could be completely incorrect.

The above situations are depicted in Figs. 7.1 and 7.2 for the circuit in Fig. 7.1(a), a simple electrical inverter and pass transistor driving a logic inverter. The correct waveforms are shown in Fig. 7.1(b). The input begins at $\mathbf{V_{OL}}$, rises to a value between $\mathbf{V_{IL}}$ and $\mathbf{V_{IH}}$ and then falls back to $\mathbf{V_{OL}}$. Correspondingly, the output falls to some low voltage and then rises back to a valid high voltage. The results from the use of different strategies to handle this case are shown in Fig. 7.2. In Fig. 7.2(a), the output of the logic gate is set to **X** since the input has stabilized at a midrange value. In Fig. 7.2(b), $\mathbf{V_{IL}}$=2.3V and so the logic output switches to $\mathbf{V_{OL}}$, but remains there since the input never crosses $\mathbf{V_{IH}}$ during its downward transition. In Fig. 7.2(c), $\mathbf{V_{IL}}$=3.0V which does not trigger the inverter at all; hence, the logic output remains at $\mathbf{V_{OH}}$ in this interval. None of these

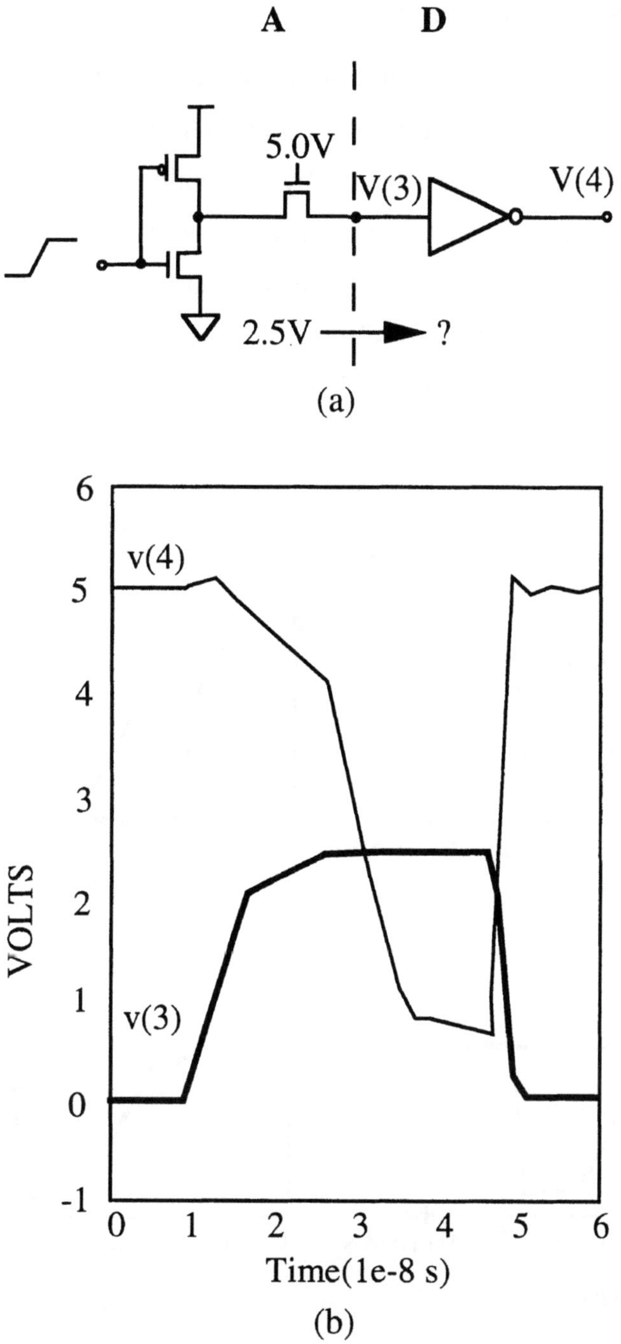

Figure 7.1: Example Circuit for A/D Interface Issues

results are particularly appealing.

There are two possible solutions to this problem. One method is to use a *dynamic mixed-mode* approach [OVE89A, THA92]. The basic idea is to replace any logic gates driven by an analog waveform by a more detailed model and to simulate it at a lower level using either timing simulation or electrical simulation. This replacement process is performed during the simulation when the possibility of generating an **X** state arises. In the case of Fig. 7.1(a), a transistor level description of the second inverter

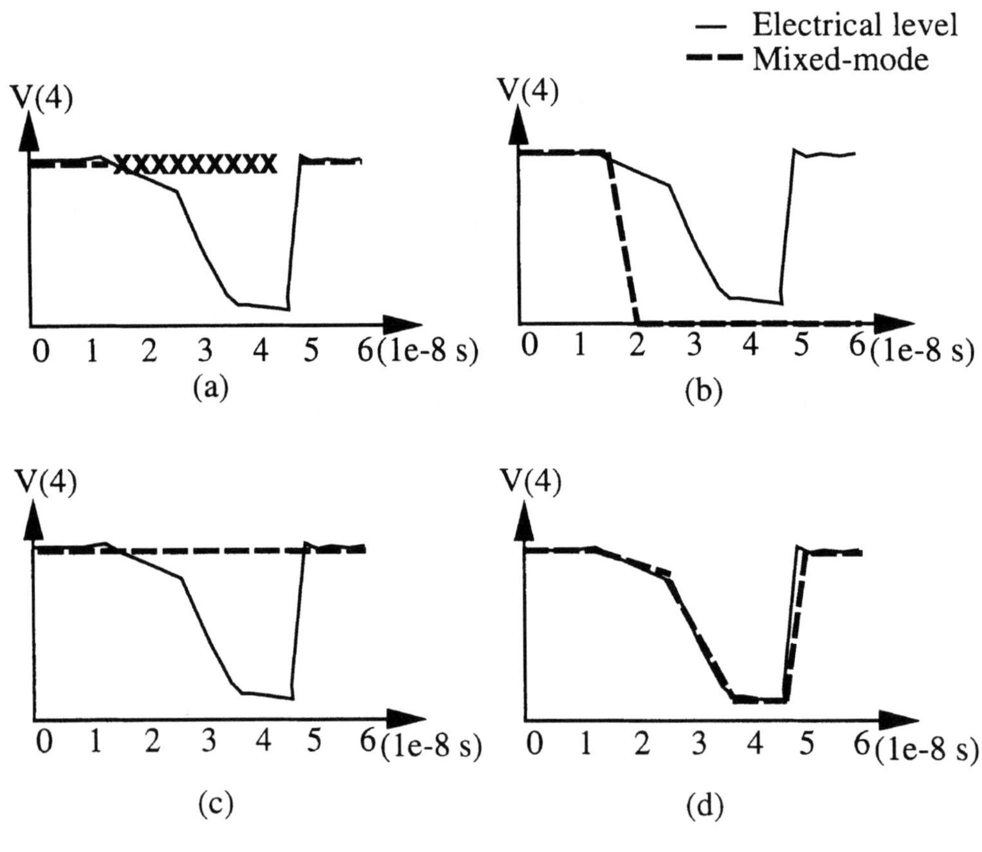

Figure 7.2: Output Waveforms for Different Approaches to handle A/D interface

would be invoked and a dc solution carried out. The simulation would continue at the electrical or switch level until a valid logic level is reached. At that point, the transistors would revert to their corresponding logic gate allowing for more efficient simulation in the rest of the interval. The main problem with this approach is its expense, and the memory requirements for storage of the circuit at different levels of abstraction. In addition, the program complexity increases greatly and this has a significant impact on maintenance and code manageability.

A more practical solution to this problem is to make use of "smart" logic gates at the analog to digital interface. Smart gates have two features that differentiate them from standard logic gates that are embedded inside a logic circuit. First, they constantly monitor the analog input for any changes in value that may affect their outputs. They ensure that the trajectory of the input is consistent with the output. If so, no change is made at the output. However, if the input switches direction, the gate recomputes its output and forces it in the opposite direction. Second, they have a more accurate voltage transfer characteristic than the standard gates. If the output stabilizes at an intermediate voltage, the output is estimated using the voltage transfer characteristic.

Fig. 7.3 illustrates two possible VTC curves that may be used in this context. The linear approximation of Fig. 7.3(a) is faster but less accurate while the cubic spline approximation of Fig. 7.3(b) better represents the true characteristic at a higher computational cost. As the approximate VTC approaches the accuracy of the actual VTC, the logic gate becomes more and more like a nonlinear macromodel, as used in [OVE88]. The simulation result from the use of a spline-fitted VTC is depicted in Fig. 7.2(d). Clearly, this solution is preferable to the other ones shown in Fig. 7.2, and the cost of this approach is only slightly higher than the basic logic gate approach [OVE89A, THA92].

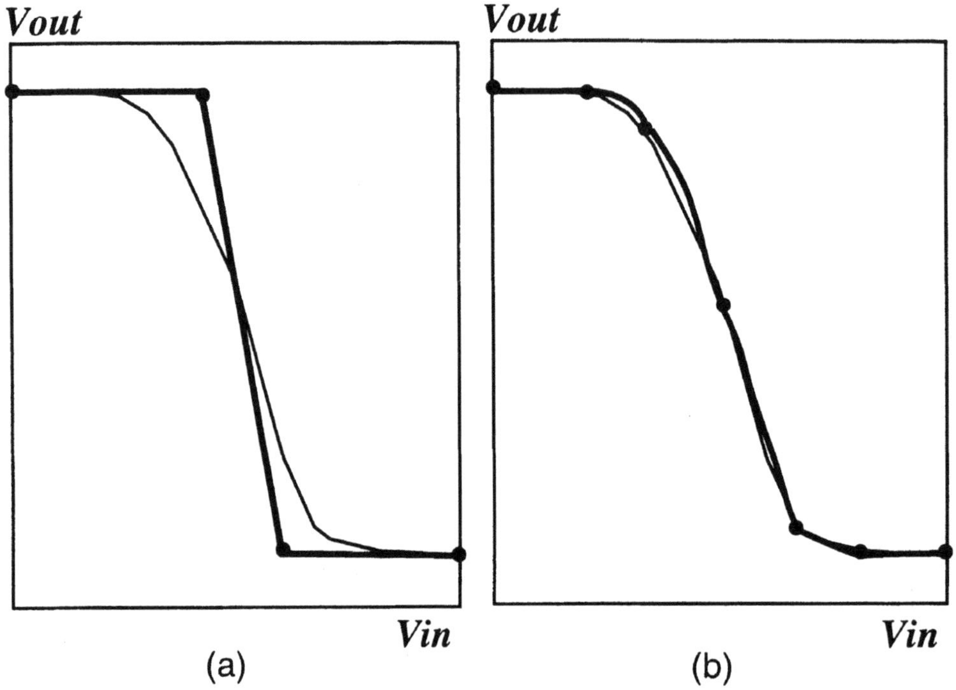

Figure 7.3: Voltage Transfer Characteristics Used in Smart Logic Gates (a) Linear Approximation (b) Cubic Spline Approximation

7.2. DIGITAL TO ANALOG INTERFACE

The translation of digital signal values to analog ones is more complicated due to the fact that information must be added at the interface in order to ensure accuracy. Early mixed-mode simulators used elements called logic-to-voltage (LTV) converters and logic-to-current (LTI) converters [NEW78] to perform signal mapping across this interface. LTV converters were used to translate logic signals that were either 0 or 1 to an equivalent electrical voltage. A finite transition time was added for rising or falling logic waveforms to avoid convergence problems in the electrical algorithms. Because the input resistance of an ideal LTV converter is zero, this model was only adequate for driving high impedance loads, such as the gate node of an MOS transistor. The LTI converter was used at the interface whenever it was necessary to model the current-sourcing or current-sinking properties of a logic gate, such as the base of a bipolar transistor. An ideal LTI converter has an infinite input impedance and is, therefore, suited to driving low impedance loads. These two converters are illustrated in Fig. 7.4.

The DIANA program [ARN78] introduced the concept of the Boolean-controlled switch (BCS) model where the state of a logic element was used to select one of two linear equivalent models. This was an

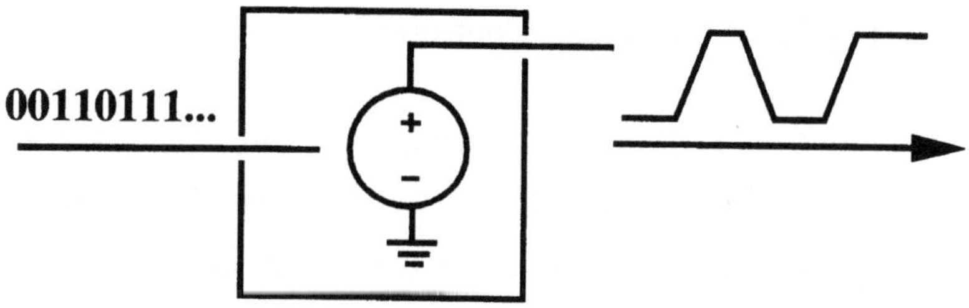

LTV:Logic-to-Voltage Converter:

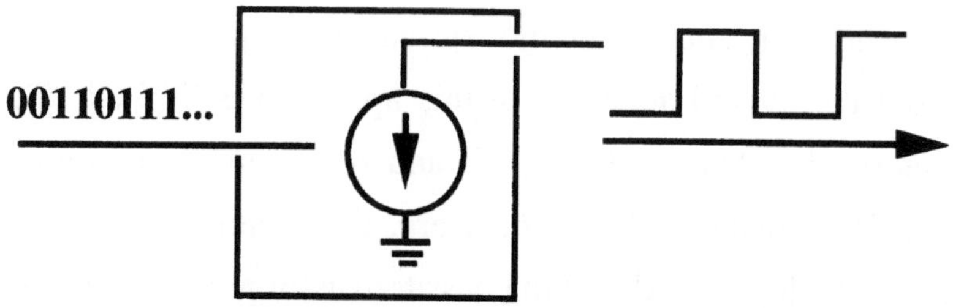

LTI:Logic-to-Current Converter:

Figure 7.4: LTV and LTI Converters

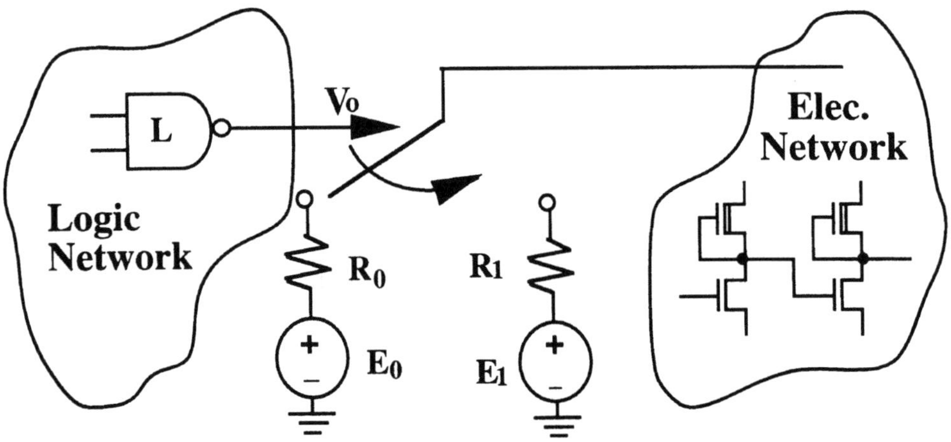

Figure 7.5: Boolean-Controlled Switch Model

important evolutionary step in the modeling of the logic to electrical interface. The BCS model depicted in Fig. 7.5 selects the **R**0–**E**0 model if the output is falling and the **R**1–**E**1 model if the output is rising. The chosen model is presented to the electrical portion of the circuit and the node is then processed as part of electrical simulation. The values of the elements in the two Thevenin equivalent models can be adjusted to improve the accuracy, but the overall accuracy of this approach is limited. To understand the reason for this, the LTV, LTI and BCS converters are all shown in the same **R–V** plane in Fig. 7.6. Here, the inverter is assumed to be represented at the logic level and the pass transistor at the electrical level.

The LTV model appears as a line at the zero resistance level while the LTI model appears as a line at the infinite resistance level. Clearly, these two models do not adequately represent the dc output characteristics

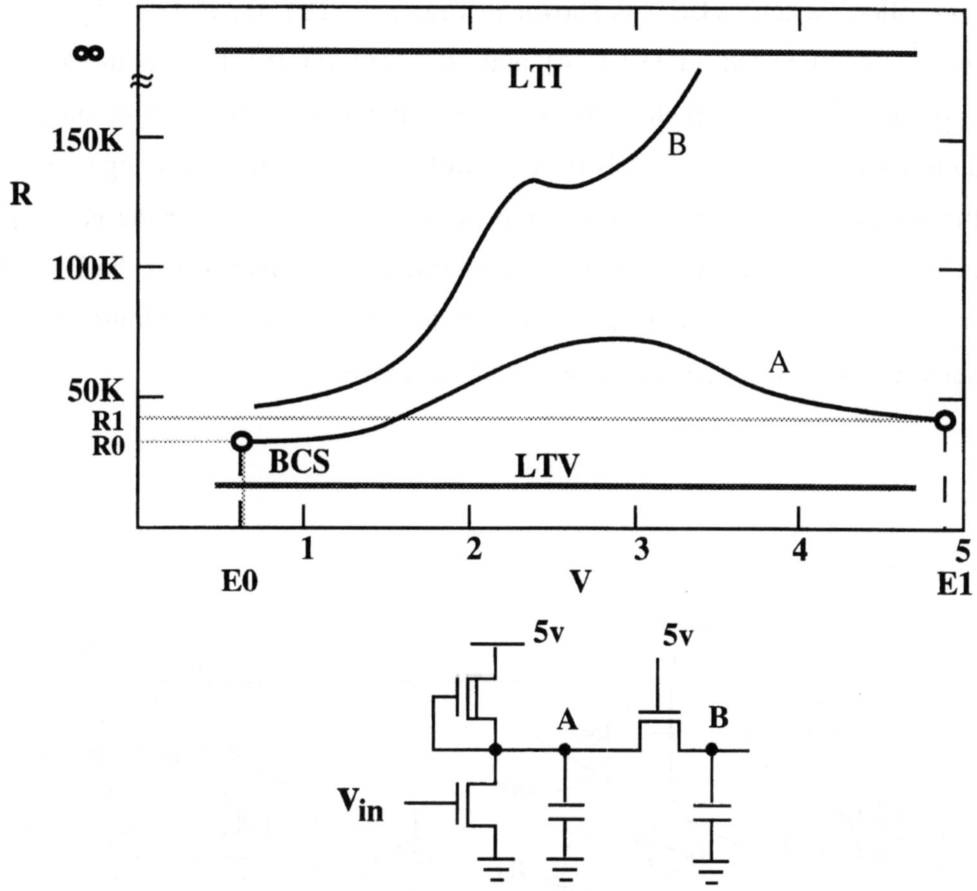

Figure 7.6: LTV, LTI and BCS in the **R**–**V** Plane

of the inverter and the accuracy during transient analysis is expected to be poor. On the other hand, the BCS model appears as two points and allows more precision in following the transfer curve. Here, the BCS mode has been chosen for the inverter output only. Unfortunately, this model only provides limited accuracy during transient analysis since the features of the curve cannot be captured by two points alone.

One way to improve the accuracy of this approach is to allow more points along the trajectory of the dc transfer characteristic. This is the basic idea behind a generalization of the BCS model, called the voltage-controlled switch (VCS), as shown in Fig. 7.7. The use of this approach was first suggested in [KLE84] and is based on the Elogic modeling approach. Rather than choosing 1-out-of-2 models to represent the output, a choice of 1-out-of-**n** models is now available. The value of **n** depends on the number of voltage levels selected by the user. The greater the value of **n**, the better the accuracy. However, as shown in Chapter 6 for the Elogic method, the CPU-time is proportional to the number of voltage states selected. Hence, a speed/accuracy tradeoff exists.

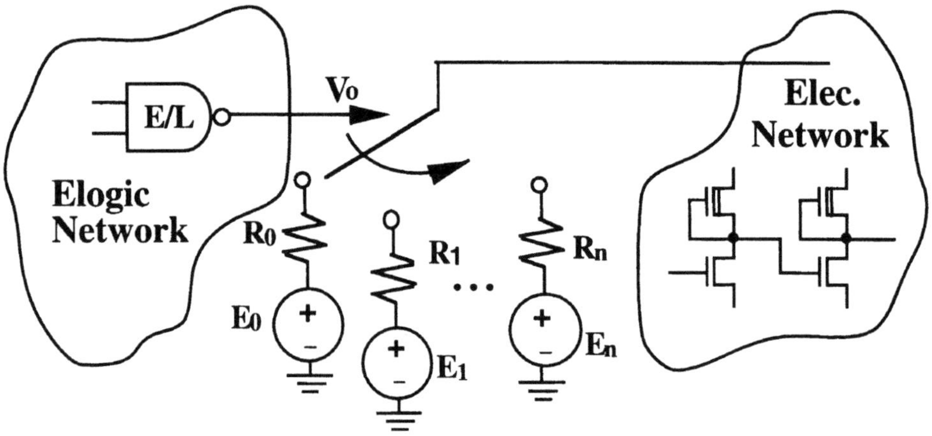

Figure 7.7: Voltage-Controlled Switch Model

The generalized model is plotted on the **R–V** plane in Fig. 7.8. A VCS model constructed from a five-state Elogic model is able to follow the trajectory of the curve quite closely; this greatly improves the accuracy during transient analysis. To better illustrate the accuracy, the two models are compared using a 1-bit memory circuit shown in Fig. 7.9. The circuit features a pair of cross-coupled inverters at the digital level interacting with a pair of bidirectional pass transistors at the electrical level. When writing data into the cell, the outputs of the inverters are modified by the incoming data and this will only occur if the output resistances of the inverters are properly taken into account. Otherwise, the stored value in the cell can not be changed. The waveform at the output of one of the inverters is plotted in Fig. 7.10 which compares the electrical solution with the BCS and VCS solutions. In Fig. 7.10(a), a BCS model is used at the

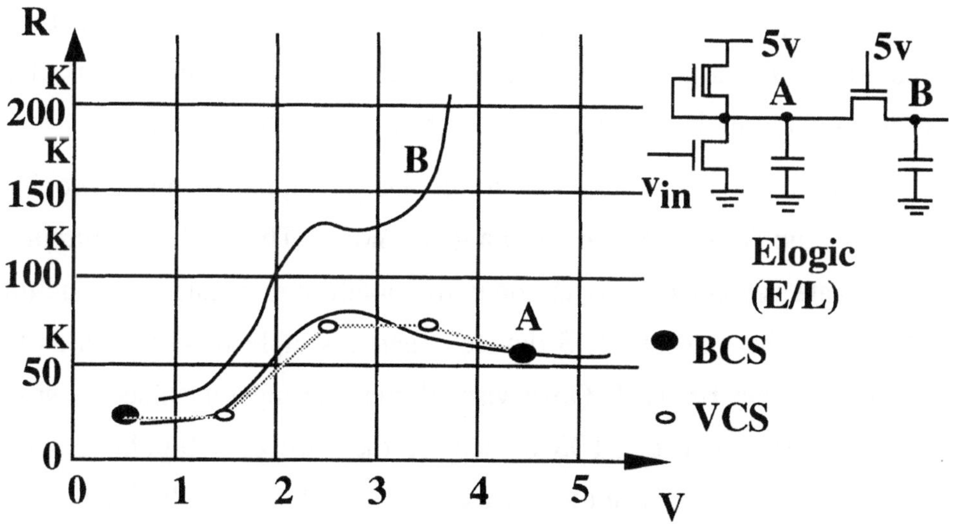

Figure 7.8: VCS Model on **R–V** Plane

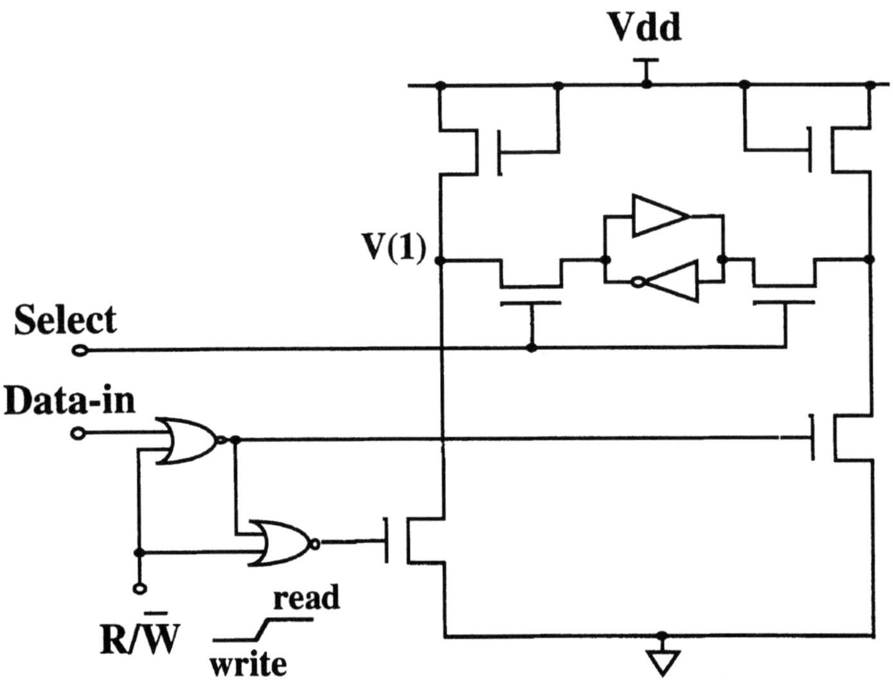

Figure 7.9: 1-bit Memory Cell

output of the logic gates and in Fig. 7.10(b), a five-point VCS model is used. Although both models produce waveforms that are essentially correct, there is a larger timing error when BCS model is used. It is clear that the VCS model is superior to the BCS model in this example.

To summarize, a general rule to resolve the signal mapping problem is based on the device connected to a given node. That is, if the output of the logic gate is connected to a node with a loading resistance approaching infinity (e.g., the gate of MOS device), then a simple model such as the LTV or BCS can be used. Otherwise, if the loading resistance is small (e.g., the drain/source of MOS device, or the base of a bipolar device), then the logic gate must use more a accurate model such as the VCS.

One other issue arises in a number of commercial mixed-mode simulators with logic simulators that represent their states as 0, 1, **X** and **Z**. As explained in Chapter 5, under certain conditions an **X** state may be

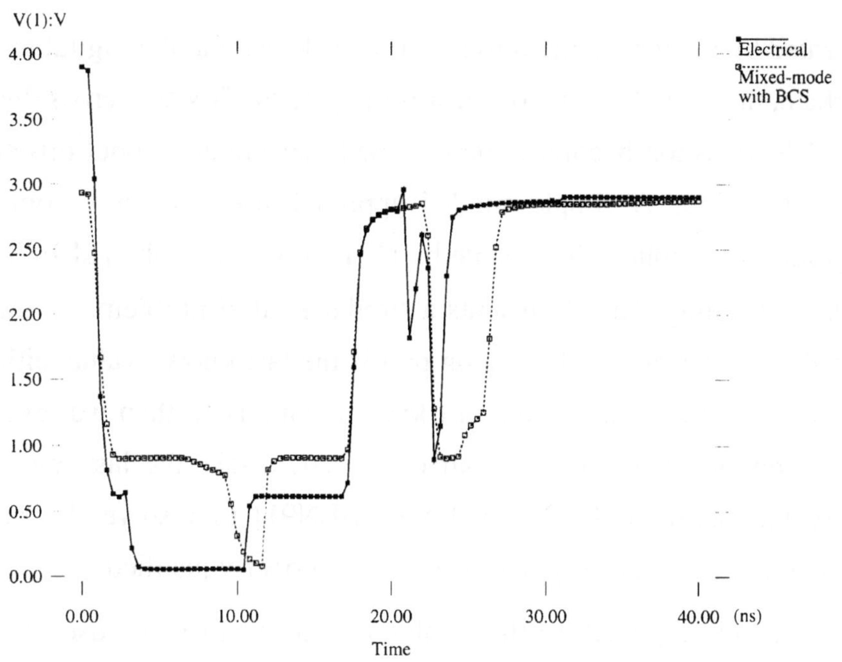

(a)

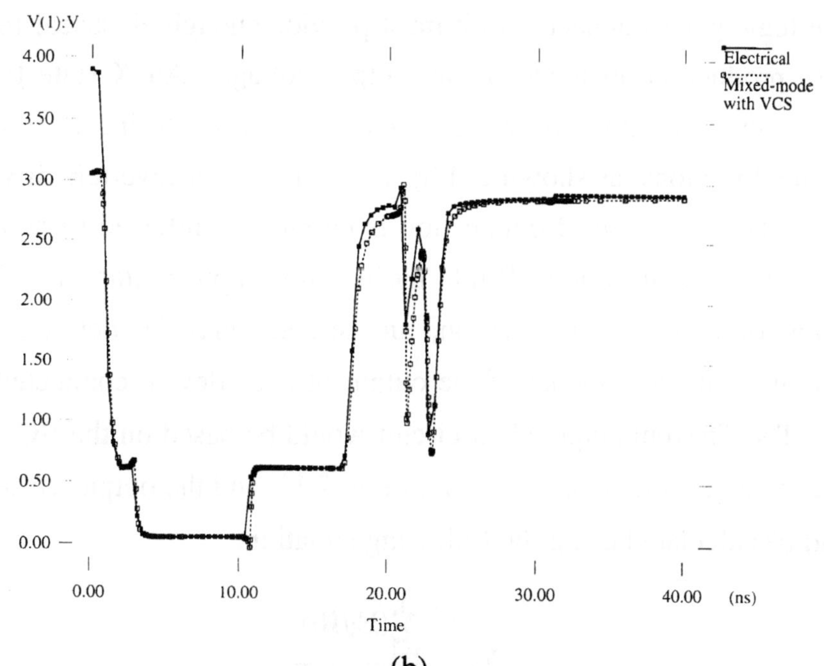

(b)

Figure 7.10: (a) Simulation of 1-bit Memory Cell using BCS
(b) Simulation of 1-bit Memory Cell using five-point VCS

generated by the logic simulation. If this is the case at the digital to analog interface, the digital simulator must propagate the **X** state across this interface. The question becomes one of how to accurately model this state on the analog side. A simple-minded approach used in many commercial programs is to replace the **X** state by $(\mathbf{V_{OH}} + \mathbf{V_{OL}})/2$ volts [LEE88]. The Z state (high impedance) presents a similar kind of problem. In this case, the value propagated to the analog part is the last known value, either 0 or 1, from the digital part. If the last known value is **X**, then we revert back to the previous problem. Of course, one could take the last logic 1 or 0 before the onset of the **X** condition [BEN91] to resolve the problem. However, none of these approaches can be properly justified.

A better approach to this problem is to again make use of "smart" logic gates. Whenever the output state of a logic node is about to become **X**, the logic gates connected to it must provide enough electrical information to produce an estimate of the output voltage. An **X** state typically arises when there are two or more drivers at a node trying to force it in opposite directions, as shown in Fig. 7.11 for a set of inverters driving the same node. This type of connection is commonly referred to as a *wired node* or *wired connection*. While the precise output voltage requires the solution of a nonlinear equation, an estimate may be derived from a Thevenin equivalent model of the output of each device connected to the node. The Thevenin equivalent circuit would be based on the BCS model of the logic gate outputs, as shown in Fig. 7.11, and the output voltage, $\mathbf{V_o}$, would be calculated using the following equation:

$$\mathbf{V_o} = \frac{\sum_{i=1}^{n} (\mathbf{E_i}/\mathbf{R_i})}{\sum_{i=1}^{n} (1/\mathbf{R_i})}$$

The value computed by this equation would be propagated to the analog

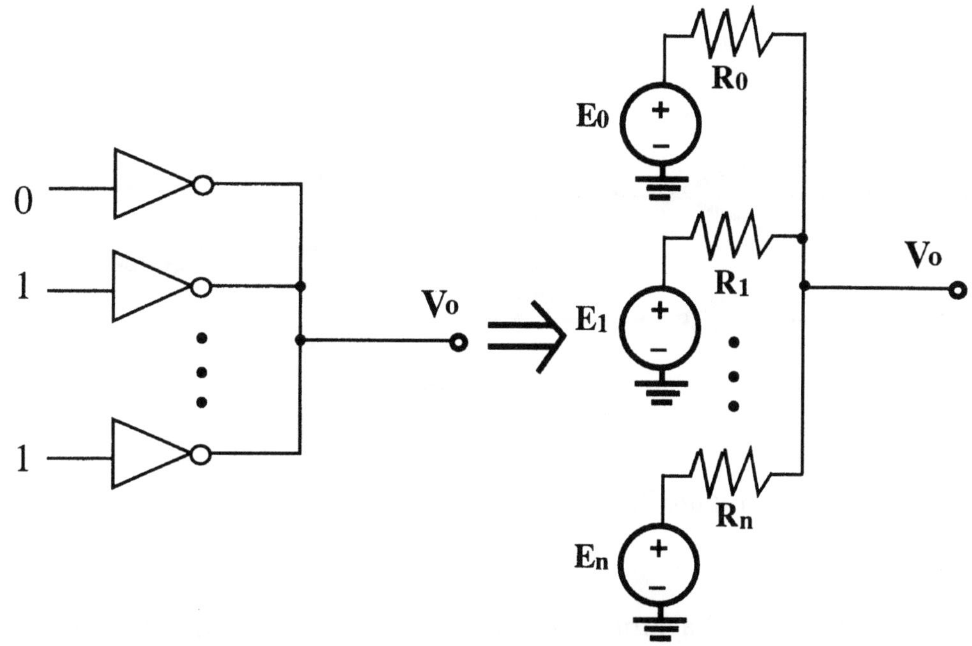

Figure 7.11: Multiple Drivers at a Logic Output

circuit across the interface. If the wired connection is not at the digital-analog boundary, then all fanout logic gates would need to make use of linear or spline-based VTC to accurately compute their outputs. This would continue until a valid high or low voltage is reached.

Since commercial logic simulators do not usually include the BCS modeling of the logic outputs, all wired connections would have to exist on the analog side of the circuit to ensure accuracy in the results. However, if the wired connection is deep within the logic circuit, this presents a problem in defining the digital-analog boundary. This is why the simple-minded approach of converting **X** to ($\mathbf{V_{OH}} + \mathbf{V_{OL}}$)/2 volts is used.

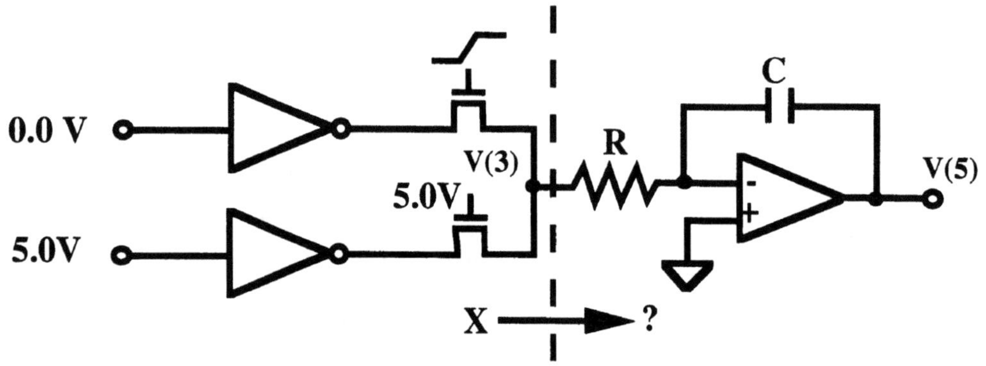

Figure 7.12: CMOS Mux-Integrator

The schematic diagram of a somewhat contrived circuit to examine **X** state mapping is shown in Fig. 7.12. Here, a MOS multiplexer is used to drive an analog integrator. The opamp is modeled using a simple voltage-controlled voltage source. The wired connection in logic domain creates a logic **X** state when both pass transistors are on. This signal is then transferred to the analog domain and into the integrator which produces a ramp function. The slope of the ramp depends on the input voltage so any error in the mapping function will be magnified by this circuit.

The results of three different simulations are shown in Fig. 7.13. The electrical simulation result produces the ramp with the least steep slope. A mixed-mode simulation using a BCS model at the interface produces a ramp that is close to the electrical solution. Also, shown are the simulation results when the **X** state is converted to 2.5V. For this case, a large error is observed in the results. Therefore, this type of **X** state mapping should be used with extreme caution, especially in critical portions of the analog blocks.

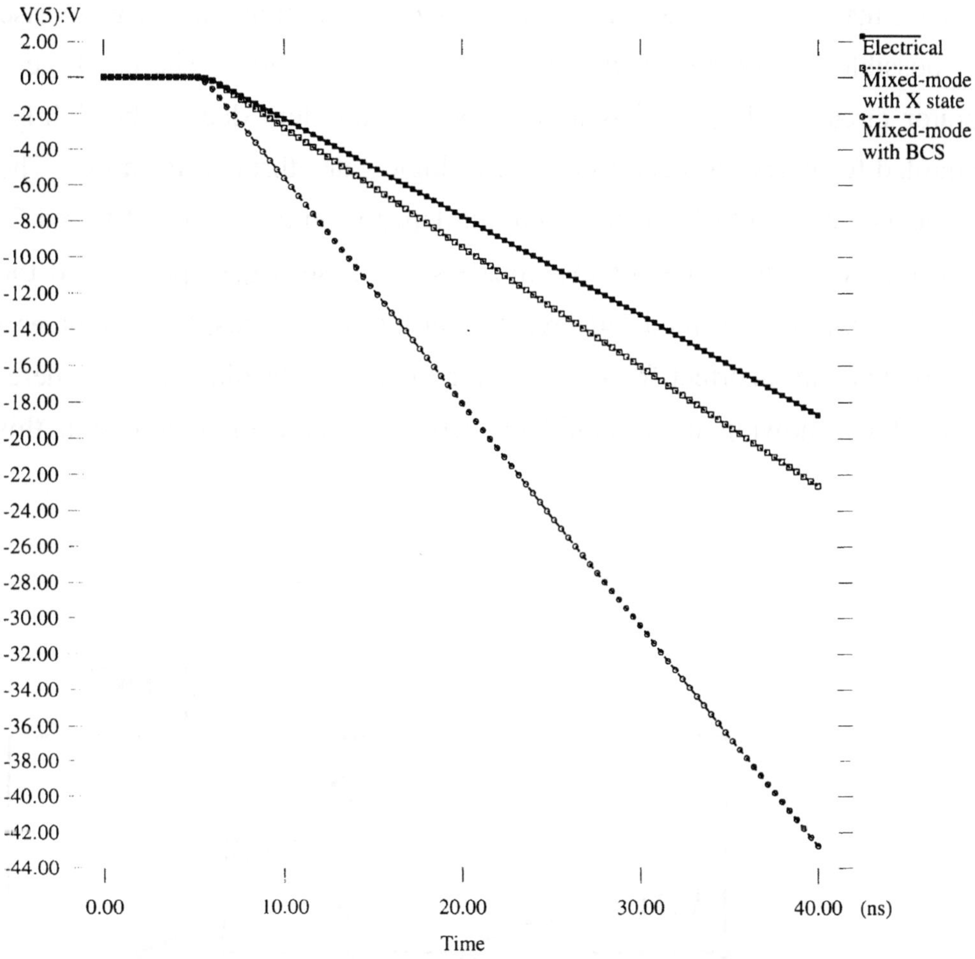

Figure 7.13: Simulation Results for Mux-Integrator

7.3. MIXED-MODE INTERFACE TEST CIRCUITS

Two additional circuits that probe the mixed-mode interface are described in this section. One of these circuits is CMOS and the other one is BiCMOS.

1) Clock Generator

Tight coupling between analog and digital components has always been a major issue for accurate mixed-mode simulation since it gives rise to possible convergence and time step control problems. The clock generator circuit in Fig. 7.14 is a common test problem due to the tightly-coupled feedback loop between the two logic gates through R and C. The frequency of the clock generator circuit depends on the choice of R and C, and the switching points of the inverters. If these trigger points and the logic output are not modeled precisely in the logic gates, the circuit will oscillate at an incorrect frequency or may not even oscillate at all. Therefore, the following tests should be performed in conjunction with this

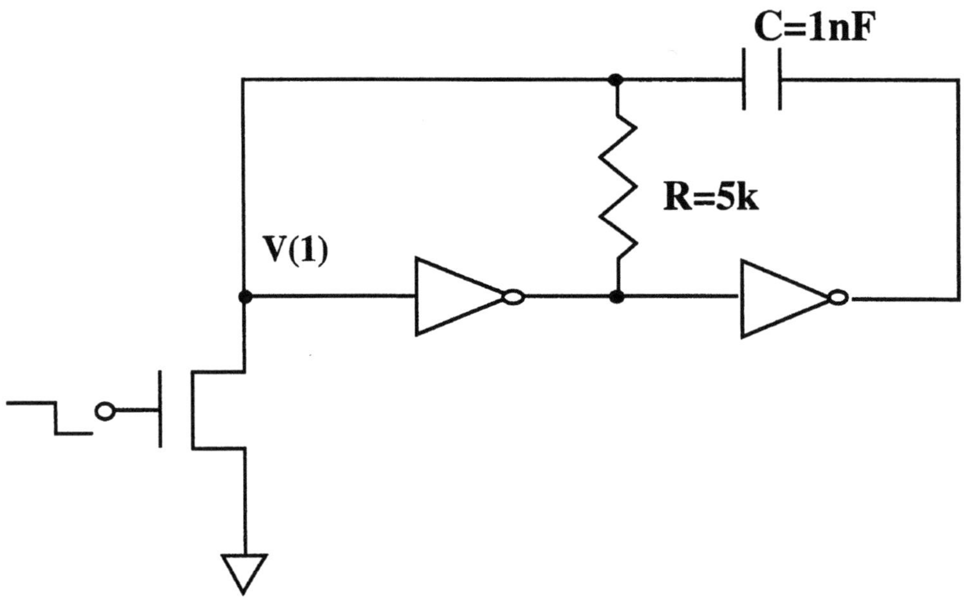

Figure 7.14: Clock Generator

circuit:

a) Adjust the trigger points of the digital inverters in the mixed-mode circuit so that the frequency generated by mixed-mode simulation and electrical simulation match each other.

b) Change R and C in both the mixed-mode and electrical circuits to obtain a different frequency. Check if the frequency generated by mixed-mode and the electrical mode still match each other.

The waveforms shown in Fig. 7.15 indicate some problems that might occur in the mixed-mode simulation. If the trigger point of logic inverter and the output equivalent model is not correct, the oscillation frequency will be different as shown in Fig. 7.15(a). After fine tuning the trigger points of the digital inverter, the two waveforms match each other as shown in Fig. 7.15(b). But if the R and C values of the circuit are changed, the results may still have some difference as shown in Fig. 7.15(c). Here, the value of R used is 10k instead of 5k. Ideally, the two sets of waveforms should be almost identical regardless of the values of R and C used.

2) BiCMOS D-Flip-Flop

The interaction of logic gates with bipolar transistors is investigated using the circuit shown in Fig. 7.16. In this case, all inverters and transmission gates are modeled in the digital domain. Only the bipolar transistors are modeled in the analog domain. Note that logic gates drive the bases of bipolar transistors, and that the emitters of the transistors are connected to the outputs of other logic gates. The modeling of the inverters driving the base of the bipolar transistors must incorporate BJT loading effects. The coupling of the feedback inverters to the BJT emitters examines the interaction of digital and analog output nodes. In effect, this

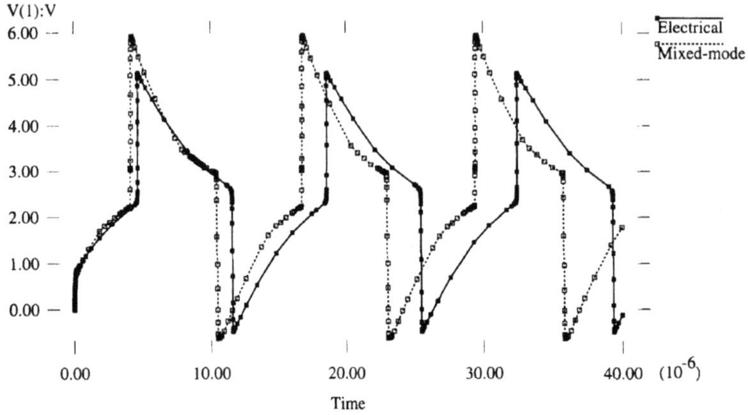

(a)

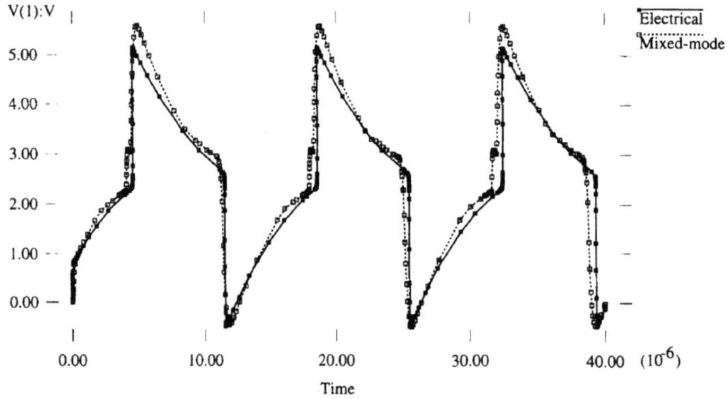

(b)

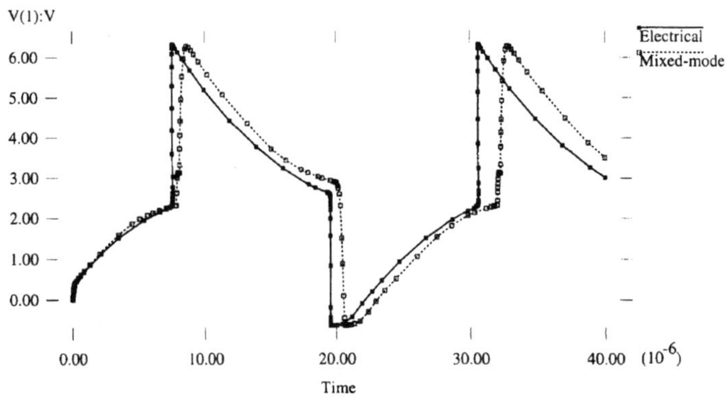

(c)

Figure 7.15: Three Test Cases for Clock Generator

circuit addresses the implementation of the digital/analog interface in both directions thereby providing a meaningful test case for bipolar mixed-signal circuits.

During the simulation of this circuit, $\mathbf{V_{IH}}$=3.2**V** and $\mathbf{V_{IL}}$=1.8**V** are used initially in the mixed-mode simulation. The waveforms shown in Fig. 7.17(a) are produced indicating that the function is incorrect. On the other hand, if $\mathbf{V_{IH}}$=2.8**V** and $\mathbf{V_{IL}}$=2.5**V** are used, and the resistance in the BCS model is fine-tuned, the function is correct but there is a noticeable timing error as shown in Fig. 7.17(b). In this case, due to the small equivalent resistance looking into the emitter of the BJTs that are connected to the local feedback inverter pairs, the voltage waveforms at these

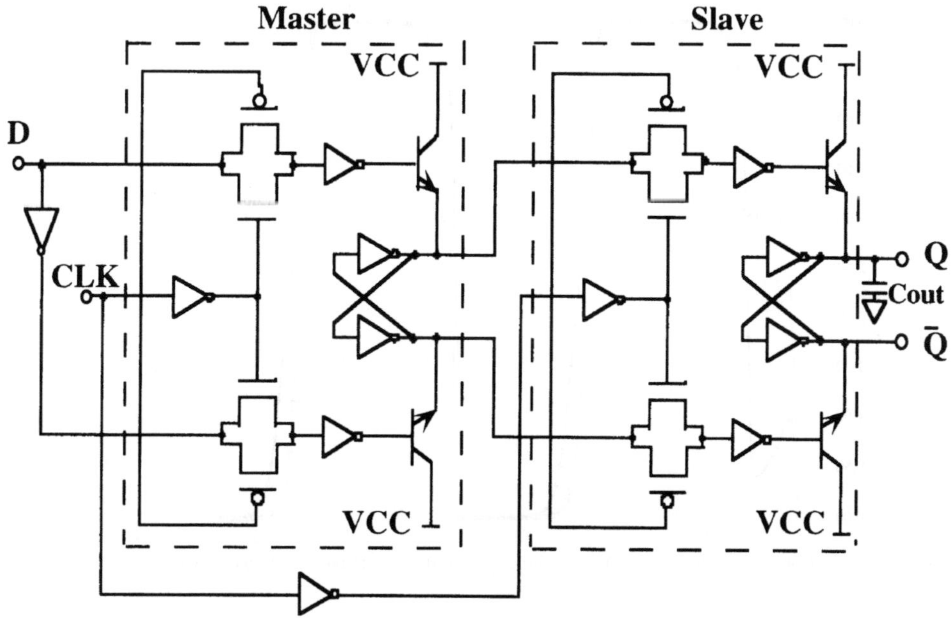

Figure 7.16: BiCMOS D-Flip-Flop

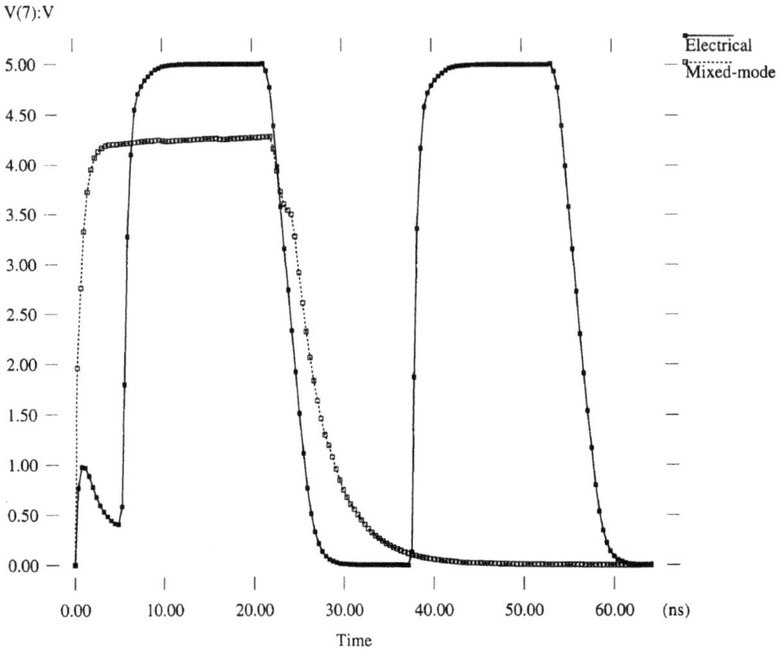

(a)

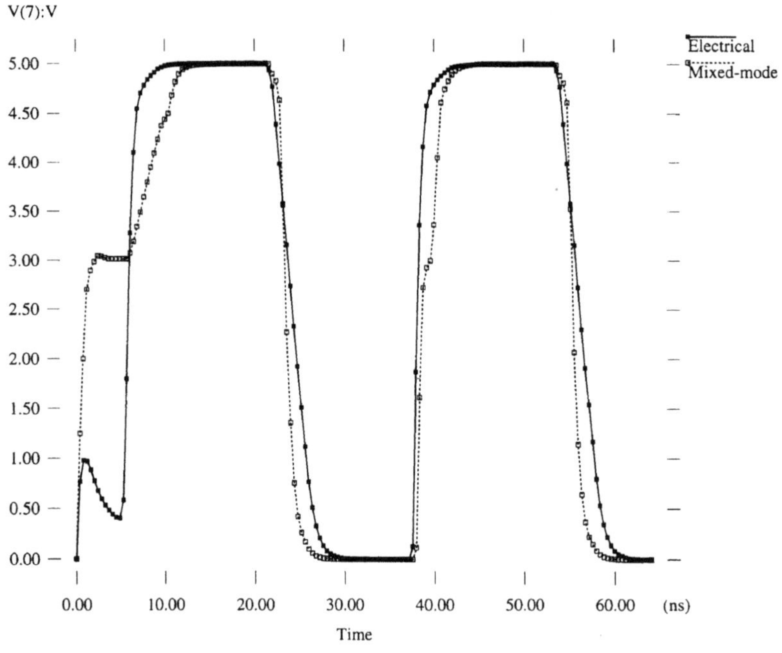

(b)

Figure 7.17: Simulation Results of BiCMOS D-Flip-Flop

nodes are very sensitive to the VTC of the digital inverters. This example is a particularly difficult to simulate accurately and is therefore an excellent test circuit for the analog/digital interface.

CHAPTER 8

MIXED-MODE SIMULATION AND IMPLEMENTATION

This chapter describes the methods involved in implementing a mixed-mode simulator and a tool that is to be used in conjunction with it. The iSPLICE3 program [SAL89A] is used as an implementation case study since it uses many of the algorithms described in the previous four chapters. The chapter begins with an overview of the architecture of iSPLICE3. Then, the mechanisms associated with the implementation of an event scheduler are presented. Following this, event scheduling policies during the transient analysis are described. Next, the techniques used to obtain the dc solution are provided. This is followed by a description of an automatic mixed-mode partitioning tool called iSPLIT [THA92]. This program converts a transistor level description into a mixed-mode description that can be used to drive the iSPLICE3 program. Finally, a large benchmark circuit is used to demonstrate the typical performance of mixed-mode simulators at the end of the chapter.

8.1. SIMULATOR ARCHITECTURE

To remain useful over its lifetime, a simulator must have the ability to adapt and grow as the technology and simulation requirements evolve. To accomplish this, a simulator should be organized so that new algorithms and models can be easily added to the existing environment. Ideally, the addition of new algorithms or models should only involve a recompilation of the program to include the newly developed routines. However, in practice, usually a few tables in a number of files must be modified to provide key pieces of information regarding the new models and algorithms.

We now describe a simulator organization that makes it relatively simple to add new simulation algorithms. In mixed-mode simulation, the use of event-driven, selective-trace in all modes of simulation is a unifying mechanism. To establish event-driven, selective-trace simulation, a time-queue and an event scheduler are required, and the notion of an *event* must be defined at each level of simulation. In iSPLICE3, each event data structure has fields for the *function*, *time* and *data* associated with the event. When an event is processed, the *function* is performed on the *data* at the prescribed *time*. New events may be scheduled in the queue as part of the call to the function. Special simulation related tasks may also be scheduled in the time queue along with regular simulation events.

The basic simulation flow of iSPLICE3 is as follows:

```
main( )
{       readin( ); build_subcircuits();
        schedule ( setup_dc_analysis event at t=0- );
        forall ( subcircuits Si in the circuit)
            schedule ( Si , t=0 ); /* for dc solution */
        schedule ( start_transient event , t=0+ );
        /* MAIN SCHEDULER LOOP: */
        while (time queue is not empty) {
            event ← GetNextEvent();
            function ← event.simulation_Mode;
            time ← event.time;
            data ← event.simulation_Data;
            /* Perform task associated with event */
            function ( data, time );
        }
}
```

□

The circuit is first read in and divided into subcircuits during the *readin()* and *build_subcircuits()* phases. The subcircuit types may be either LOGIC, ELOGIC, or ELECTRICAL depending on the node type. iSPLICE3 determines the node type based on the devices connected to nodes in the circuit. For example, if a node has only ELOGIC devices connected to it, it will be labeled as an ELOGIC node. If it has only LOGIC devices connected to it, then it will be labeled as a LOGIC node. However, if there is at least one ELOGIC device controlling the node (i.e., the drain or source of a ELOGIC transistor), it will be labeled as an ELOGIC node. Similarly, if there is at least one ELECTRICAL device with a controlling node connected to it, it is labeled as an ELECTRICAL node. After the node assignments are completed, the ELECTRICAL nodes are further partitioned into subcircuits of tightly-coupled nodes as part of the standard ITA relaxation algorithm. Finally, the subcircuits and fanin and fanout tables are constructed using the node assignment information.

The next step is to schedule the *setup_dc_analysis()* event, and then schedule all the newly created subcircuits for evaluation at time t=0 as part of the dc solution. The last event to be scheduled before entering the processing loop is the *start_transient()* event, which is executed immediately after the dc solution is obtained. The program then enters the main loop where the scheduler sequences through the list of scheduled events. It remains in this loop until there are no events in the queue, at which time the program stops. The inner part of the loop involves obtaining the next event from the queue and then executing the function associated with the event. Examples of simulation functions that may be scheduled in the time queue are *ELECTRICAL_event()*, *LOGIC_event()* and *ELOGIC_event()*.

The basic flow for a simulation event is shown below:

```
simulation_event( Si, tN )
{
        get_input_voltages(Si );
        process_subcircuit(Si);
        if (Si is active)
            schedule ( Si , tN+1 );
        foreach ( node j in Si)
            foreach ( fanout subcircuit Sj of node j )
                if (node j has crossed a critical threshold of Sj)
                    schedule ( Sj , tN );
}
```

□

First the external voltages for the subcircuit are obtained. Then the subcircuit is processed using the appropriate analysis mode. If the voltages in the subcircuit have changed, the subcircuit is rescheduled for evaluation at a later time. Then the fanouts are scheduled at the current time if any important thresholds have been encountered.

8.2. EVENT SCHEDULER DESIGN

In this section, a number of alternative strategies for the implementation of event schedulers are described. In designing a scheduler, a number of important issues relating to scheduler function and efficiency must be addressed. First, the event scheduler must have some notion of a time sequence and must be able to associate an event with a particular point in time. It may also be necessary to arrange events at a particular time point in some sorted order. Occasionally, the simulator will schedule an event and later decide that the event is no longer necessary. Hence, the scheduler must have the capability of canceling a pending event. Finally, the scheduling operations must be efficient, since they add to the simulation overhead. The event insertion/deletion operations must be relatively fast and the time sequencing through events should be efficient. Both of these requirements can be achieved by maintaining some uniformity in the event distributions in the scheduler, as will be seen. The scheduler overhead is usually insignificant for electrical analysis (since the events themselves are usually computationally intensive), but it may be a dominant factor in switch-level or higher levels of simulation where event processing operations are relatively simple. In general, the scheduling overhead should not consume more than 5-10% of the total simulation time. With these considerations in mind, the event schedulers commonly found in mixed-mode simulators are presented.

The classical approach to the scheduler design uses a *time-wheel* [BRE76] mechanism as illustrated in Fig. 8.1. This structure allows the indexed list to "wrap-around" so that the array of headers can be reused once the events associated with that entry have all been processed and the **PT** pointer shown in the figure has been incremented. For example, when the events at time **PT** have all been processed, the header at **PT** can be reused to represent the time **PT+MΔt**, assuming that the array has **M** elements. Using the MOD function, the **PT** pointer is updated as follows:

$$PT = (PT + 1) \text{ MOD } M$$

The obvious advantage of this approach is that a predetermined amount of memory can be allocated for the time queue *a priori*. However, dynamic memory allocation will still be required for events which occur at time points greater than **t+MΔt** units in the future. They may be organized in an *overflow* or *remote list*. At some point, these remote events must be brought into the time queue. Since the events in the remote list are usually somewhat more expensive to insert or remove than events in the time queue, it is not efficient to update the time queue with events from the

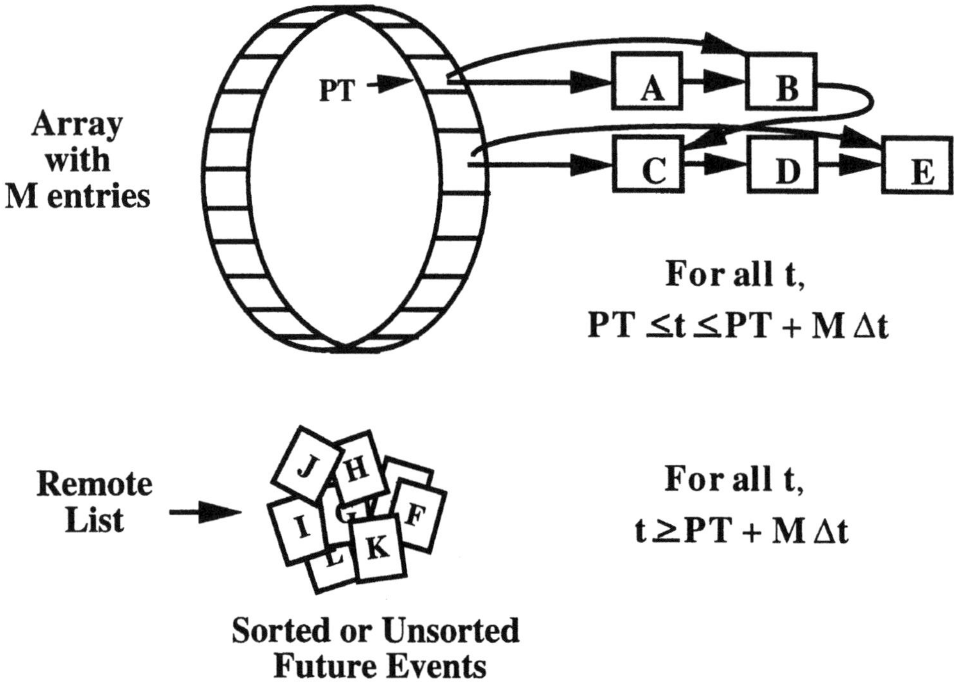

Figure 8.1: Classical Time-Wheel Mechanism

remote list every time a time-point event list has been processed. However, as more and more time points are processed, the probability that new events will end up in the remote list increases, and this is undesirable. Therefore, it is better to move events from the remote list to the time queue periodically, i.e., after processing **k** time points in the time wheel.

Another source of inefficiency is due to the fact that many headers may not point to any events. These headers must be scanned anyway and this consumes additional CPU-time. The distribution of the events in the time queue, hence the sparsity, depends on the value of **Δt**. For example, if Δt is very small, only a few events will be scheduled at each time point. On the other hand, if Δt is large, the events will densely populate the region of time near the current time pointer, **PT**. Both situations will reduce the scheduler performance. Hence, the number of time points processed before bringing in remote events (**k**), the size of the time wheel (**M**) and step size between adjacent entries in the time wheel (Δt), and indeed how efficiently the remote list is managed, all have an impact on the efficiency of this type of scheduler. Typical event distributions should be examined to select the appropriate values for these parameters for a given application.

The remote list usually contains a small number of events if the proper parameter values are selected for the time queue. It usually contains events associated with external sources and these events are often sparse in time. The objective is to ensure that the ratio between events in the time queue and remote list does not exceed a certain threshold. If it is not anticipated that many events will be scheduled in the remote list, it may be organized as a simple linear linked list.

There are other situations where a more elaborate organization of the remote queue is required. In the case of electrical simulation, some components may take small time steps during a transition while others use very large time steps due to the latency. Here, a *secondary* time wheel would be useful. It can be managed in exactly the same way as the primary time wheel except that each interval is defined to be $k\Delta t$ units of time. After $k\Delta t$ units of time have been processed in the primary time queue, all the events in the next interval of the secondary queue can be moved to the primary queue. In general, it is possible to have a set of remote time wheels, each having an interval, Δt_i, which is equal to $k\Delta t_{i-1}$, where Δt_{i-1} is the interval used by the previous time-wheel. If a variable number of time wheels are used, another level of indexing would be useful in selecting the proper time-wheel.

The scheduler used in iSPLICE3 as shown in Fig. 8.2 is similar to the classical time-wheel mechanism. However, rather than a single time-wheel, a pair of time queues with **M/2** entries and a remote list are used. While events are being processed from the first queue, new events may be scheduled either in the remaining portion of the first queue, in the second queue or in the remote list. The remote list is maintained as a simple linear linked-list. When the end of the first queue is reached, the second queue becomes active and the first queue is adjusted to represent the next (**M/**2)**Δt** units of time. Any appropriate remote events are moved from the remote list to the first queue. When the end of the second queue is reached, the first queue becomes the active queue again while the second queue is modified to represent the next (**M/**2)**Δt** units of time. This scheme represents a compromise between bringing in new events after each time point list is processed and bringing in remote events only after all events in the queue have been processed.

8.3. TRANSIENT ANALYSIS AND EVENT SCHEDULING

iSPLICE3 performs both dc analysis and time-domain, transient analysis of MOS and bipolar integrated circuits. Transient analysis is generally the most time-consuming and memory-intensive task in simulation but the mixed-mode techniques used in the iSPLICE3 program can reduce the simulation time significantly compared to that for SPICE2. iSPLICE3 has three simulation modes: circuit level simulation (ELECTRICAL) which uses iterated timing analysis, switch-level timing simulation (ELOGIC) and gate-level logic simulation (LOGIC). Each mode can be used independently or combined in a mixed-mode simulation. The details of each algorithm have been described in the previous chapters.

One issue that has been overlooked is that of event scheduling

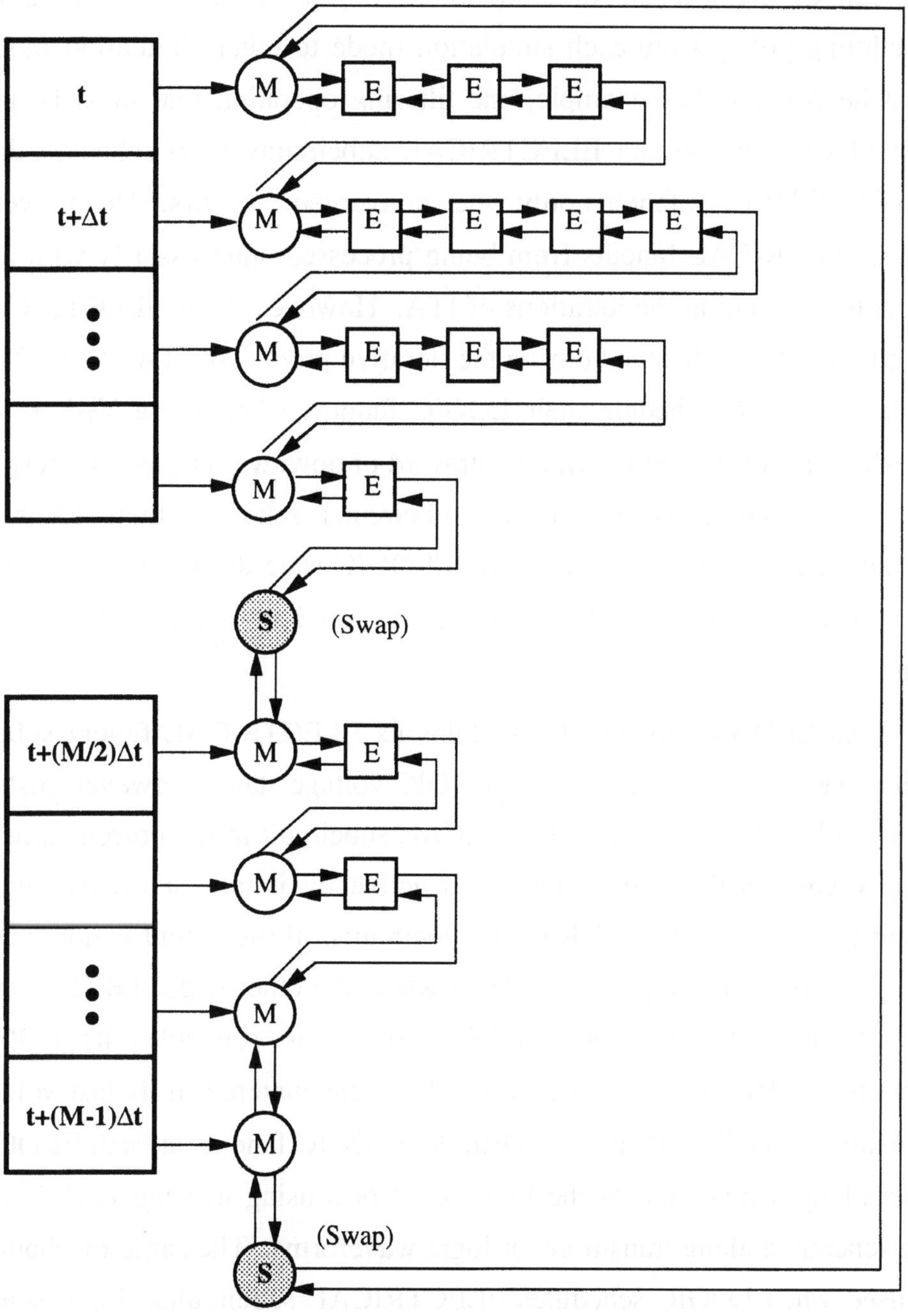

Figure 8.2: iSPLICE3 Time Queue Data Structure

between different levels of simulation. A set of rules governing the scheduling policy from each simulation mode to other simulation modes must be defined. For example, one filtering operation that must be performed when processing ELECTRICAL subcircuits is to schedule their non-ELECTRICAL fanouts only when convergence occurs. This prevents non-ELECTRICAL fanouts from being processed unnecessarily with partial solutions during the iterations of ITA. However, other ELECTRICAL fanouts must be scheduled during the iterative process of ITA. ELECTRICAL subcircuits schedule their LOGIC fanouts whenever a $\mathbf{V_{IL}}$ or $\mathbf{V_{IH}}$ threshold is encountered during an upward or downward transition, respectively. Similarly, ELECTRICAL subcircuits schedule their ELOGIC fanouts if they have encountered an ELOGIC state during the last transition. This is consistent with the scheduling used among ELOGIC subcircuits.

An ELOGIC subcircuit schedules its ELECTRICAL fanout subcircuits whenever it reaches a new ELOGIC voltage state. However, instead of actually scheduling an ELECTRICAL subcircuit at the current time, it simply ensures that the subcircuit is active by issuing a *wakeup_call()* event to any fanout ELECTRICAL subcircuits. If the fanout is not active, the *wakeup_call()* simply schedules it where the other ELECTRICAL subcircuits are scheduled. An ELOGIC subcircuit schedules its LOGIC fanouts if a $\mathbf{V_{IL}}$ or $\mathbf{V_{IH}}$ threshold has been encountered in its last voltage transition. LOGIC subcircuits schedule ELOGIC fanouts at each ELOGIC state along a transition of the logic waveform using *wakeup_call()*'s that are scheduled along transitions of logic waveforms. The same mechanism is used when LOGIC schedules ELECTRICAL subcircuits. Input source events follow similar rules as described above and are also dependent on the types of devices connected to them.

One additional complicating factor in intersimulation scheduling is due to *roll-back* or step rejections. It may be necessary to occasionally cancel a pending event or reject a time-step and begin reprocessing at an earlier time. When this occurs, the scheduler must be backed up and the subcircuits rescheduled and reprocessed accordingly. The subcircuit which encountered the rejection is processed initially. If its new solution differs significantly from the previous one, its fanout subcircuits are scheduled. Otherwise, no scheduling operations are performed. Similarly, the fanouts are processed and they compare their newly computed solutions with previous solutions and schedule their fanouts only if the new solutions are different from their old solutions. Both ELOGIC and LOGIC schedule events on fixed grid boundaries so that slight variations in the computed schedule times are not inferred as different solutions. The roll-back strategy ensures that accurate solutions will be obtained in an efficient manner.

8.4. DC ANALYSIS TECHNIQUES

iSPLICE3 provides a number of different techniques to obtain a dc solution for a given circuit. For ELECTRICAL circuits, either the standard Newton method, source-stepping or gmin_stepping methods may be invoked [QUA89]. For circuits that are represented using the ELECTRICAL, LOGIC and ELOGIC levels, iSPLICE3 uses an iterative mixed-mode dc solution scheme to initialize the node voltages, as follows:

```
dc_analysis()
{
    repeat {
        process_LOGIC_nodes(); /* using logic simulation */
        process_ELECTRICAL_and_ELOGIC_nodes();
        /* using Newton's method */
    } until (convergence)
```

```
        set_ELOGIC_nodes(); /* force to discrete values */
        repeat { /* correct any nodes affected by last operation */
            process_LOGIC_nodes();
            process_ELECTRICAL_nodes();
            /* leave out ELOGIC nodes */
        } until (convergence)
    }
```
□

The algorithm given above is performed at time 0 using event-driven techniques. First, the LOGIC nodes are processed using zero-delay logic simulation. Then the ELECTRICAL and ELOGIC nodes are processed using direct methods (i.e., the standard Newton method). Any nodes that are different from their previous solution act to schedule their fanout nodes at time 0. This process is repeated until convergence occurs. When the dc solution is obtained, the ELOGIC nodes are set to their nearest discrete values and the iterative loop is repeated once again to correct any values that may be affected by this operation. Unfortunately, the convergence of the dc solution is not guaranteed in all cases. In fact, if the LOGIC nodes do not have a dc solution, or if a proper initial guess is not specified for the ELECTRICAL and ELOGIC nodes, the iterative process may not converge at all!

While it is generally difficult to find a dc solution for LOGIC nodes that may oscillate when analyzed using zero-delay logic simulation, iSPLICE3 uses a new technique to improve the likelihood of convergence for ELECTRICAL and ELOGIC nodes in MOS digital circuits. This technique provides an initial guess that is usually close to the final solution, it ensures proper and reliable convergence and reduces the total number of Newton-Raphson iterations required. on MOS digital circuits described at the transistor level. First, the ELECTRICAL and ELOGIC portions of the circuit are solved using zero-delay, switch-level logic simulation [BRY80] to derive the initial conditions at each node. Then these logic values are converted to their corresponding voltage values. Next, the standard Newton method is applied to the same portion of the circuit, using the derived values as initial guesses. Since this technique provides an initial guess that is usually close to the final solution, it ensures proper and reliable convergence and reduces the number of overall iterations. This approach has been found to be 4-5 times faster than the standard approach on MOS digital circuits and successfully converges on circuits that fail to converge in SPICE2.

In the simple algorithm above, the processing of feedback paths

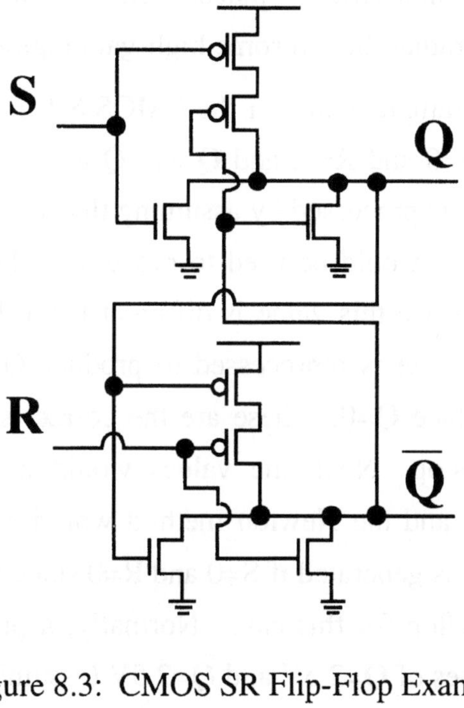

Figure 8.3: CMOS SR Flip-Flop Example

deserves some special attention since all nodes are set to the uninitialized state as the first step of the switch-level analysis at time 0. iSPLICE3 processes the nodes from the inputs to the outputs, but if there are feedback paths in the network, some of the node values needed for the evaluation may be uninitialized, which presents a problem in determining the state of the output node. For these situations, iSPLICE3 guesses the values of initial unknowns whenever required as either logic **0** or logic **1**, depending on the situation. If an incorrect guess is made, the feedback path will act to correct the situation in a subsequent processing step. This technique removes most of the uninitialized states at the output nodes, particularly in troublesome circuits such as flip-flops. However, some nodes may be assigned to the **X** state if the correct state can not be determined during switch-level simulation. These nodes are reset to 0 **V** before applying the Newton method since it places NMOS transistors in the cutoff region of operation rather than in some high-gain region.

As a simple example, consider the CMOS SR flip-flop circuit in Fig. 8.3. Assume that S=0 and R=1, and **Q** and $\overline{\mathbf{Q}}$ are uninitialized. Then, if the upper NOR gate is processed by assuming that **Q**=1, a value of $\overline{\mathbf{Q}}$=0 is produced. This value would be used to process the lower NOR gate and **Q**=0 is produced. Since this value is different from the original assumption, the first NOR gate is reprocessed to produce $\overline{\mathbf{Q}}$=1, and the second reprocessed to produce **Q**=0. These are the correct solutions and so the processing would stop. Next, the values would be converted to their equivalent voltages and the Newton method would be invoked. A more interesting example is generated if S=0 and R=0 since the previous outputs are held in the flip-flop for this case. Normally, a program like SPICE2 would produce values of **Q**=2.5V and $\overline{\mathbf{Q}}$=2.5V (assuming a 5V supply voltage) as the dc solution, which is clearly incorrect. iSPLICE3 will produce either **Q**=0.0 and $\overline{\mathbf{Q}}$=5.0 or **Q**=5.0 and $\overline{\mathbf{Q}}$=0.0 and either case is an acceptable solution. Of course, the user can always override these values by initializing the flip-flops to any desired setting.

8.5. AUTOMATIC MIXED-MODE PARTITIONING

One problem that has not been directly addressed until now is that of defining the portions of a circuit to be represented at the various levels of abstraction. Normally, this task is the responsibility of the user since the circuit designer has the knowledge to perform the operation manually. However, one tool that would prove to be extremely useful is an automatic mixed-mode partitioner. Such a tool would be necessary when a transistor level description is extracted from a layout, or obtained from any other source, and a functional verification is desired in as short a time as possible. The partitioner would scan the circuit description and define the different levels of abstraction that would be used to simulate different portions of the circuit and then provide this information to the mixed-mode simulator.

Conceptually, this process takes a collection of components from a given level in the circuit description and replaces them with higher level primitives[1] to improve speed [RAO89, ACU89], or with lower level primitives to improve accuracy [OVE89A]. The complete partitioning operation involves two phases: recognition and characterization. Groups of components that combine to form higher level primitives in the circuit must first be recognized using either a rule-based approach or a table lookup scheme. Then the parameters for the higher level model must be generated from the lower level description in the characterization phase to maintain simulation accuracy.

---

[1] A primitive refers to a basic element that is known to the simulator, i.e., any element that is hard-coded into the program.

8.5.1. Program Overview

In the following sections, a program called iSPLIT [THA92] that performs automatic partitioning for mixed-mode simulators is described. The iSPLIT program has three phases, as shown in Fig. 8.4. In the first phase, the program searches through a transistor netlist and substitutes logic models for various gates found in the netlist. This process is facilitated by defining transistors groups, based on channel-connected components, and performing a gate-recognition algorithm on each group. It can recognize any inverters, NANDs, NORs, or complex gates in CMOS, NMOS, or pseudo-NMOS technologies. In the second phase, the program reclassifies all of the groups that were not recognized above, by assigning each one a new type number. Third, any user-defined cells that are specified in a library file are replaced with their corresponding higher-level block. Any transistor groups that are not recognized after these three steps are left at the transistor level. The program then writes out a new netlist that can be used for mixed-mode simulation.

8.5.2. Channel-Connected Transistor Groups

In the first phase, all of the MOS transistors are partitioned into groups. Each group consists of transistors which are connected at their source and drain terminals. For convenience, the PMOS and NMOS transistors are kept in separate lists in the group. The list of PMOS transistors is called the p-tree, and the list of NMOS transistors is called the n-tree. The p-tree may also contain depletion NMOS transistors if the gate is an NMOS gate. During the grouping process, any node which has both PMOS and NMOS transistors connected to it (or both enhancement and depletion NMOS transistors) is considered as a possible output node. Because all logic gates and complex gates have only one output, only one output node is allowed in each group. If a second possible output node is found, the program will backtrack, and not include that node in the group.

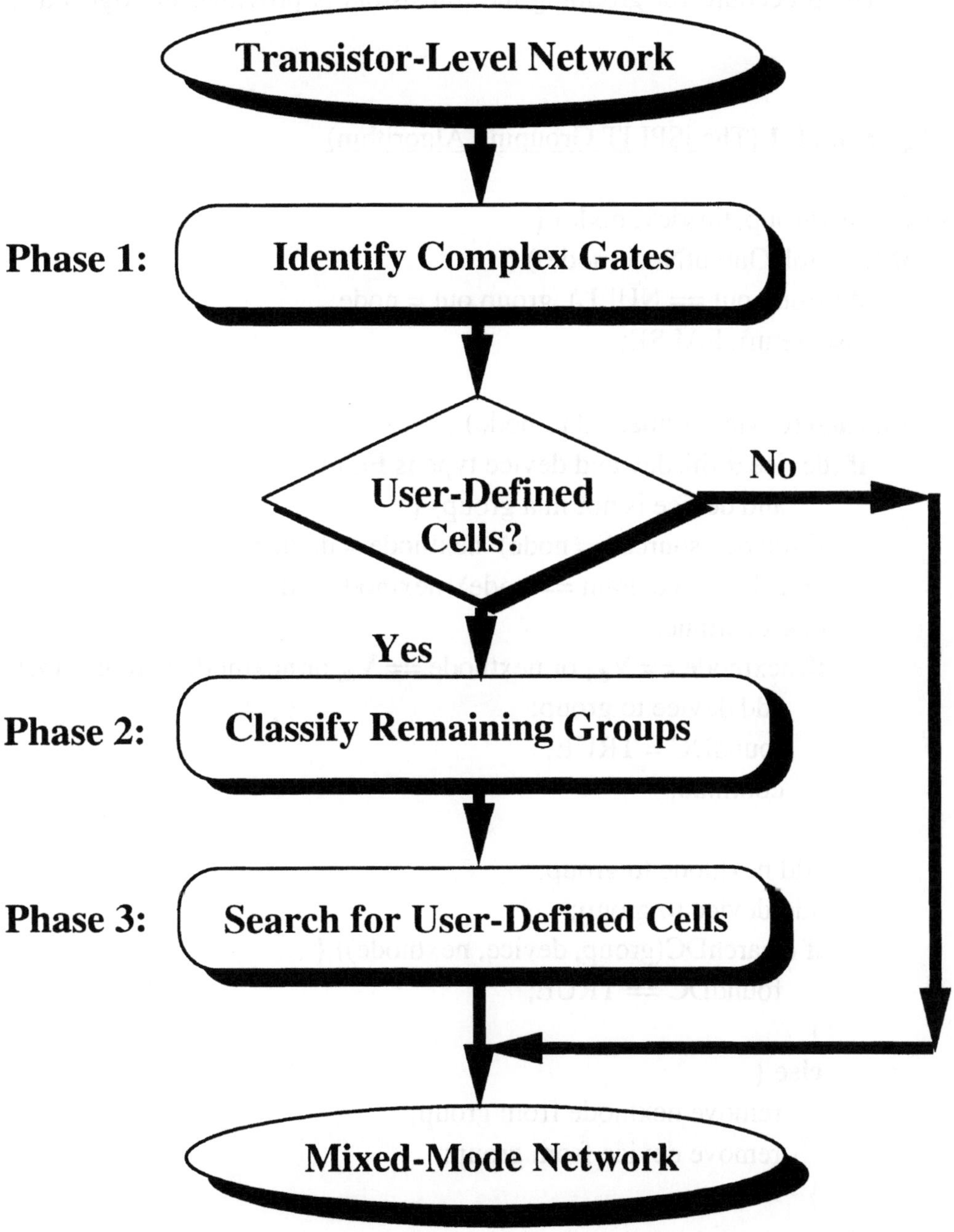

Figure 8.4: Overview of the iSPLIT Program

The procedure for grouping the transistors is provided in Algorithm 8.1.

__**Algorithm 8.1** (The iSPLIT Grouping Algorithm)__

```
searchDC(group, thisdev, node) {
    if (possibleOutputNode(node)) {
        if (group.out == NULL)  group.out = node;
        else  return FALSE;
    }
    foreach (device connected to node) {
        if (device ≠ thisdev and device type is FET1
                and device is not in a group) {
            if (device.source == node)  nextnode = device.drain;
            else if (device.drain == node)  nextnode = device.source;
            else continue;
            if(nextnode == Vdd or nextnode == Vss or nextnode is in group){
                add device to group;
                foundDC = TRUE;
                continue;
            }
            add nextnode to group;
            add device to group;
            if (searchDC(group, device, nextnode)) {
                foundDC == TRUE;
            }
            else {
                remove nextnode from group;
                remove device from group;
            }
        }
    }
    return foundDC;
}
```

□

It is a recursive, depth-first search, which examines all paths from a given node to other nodes through a transistor. The program begins by finding all PMOS or depletion NMOS transistors that are connected to $\mathbf{V_{dd}}$. Each of these transistors is considered as a starting point for a new group. When processing each device, the program checks that the device does not already belong to a group. If either the source or drain node of the transistor is connected to a node that is not $\mathbf{V_{dd}}$ or $\mathbf{V_{ss}}$, then it creates a new group and calls the *searchDC* function to find all the transistors that belong to that group.

The *searchDC* function takes a group, a transistor and a node as arguments. The function examines every transistor connected to the given node by its source or drain, except for the transistor passed into the function. For each transistor, it finds the other node to which the channel of the transistor is connected. If the present node is connected to the source, the next node is the node connected to the drain of the transistor. If the next node is either $\mathbf{V_{dd}}$, $\mathbf{V_{ss}}$, or already in the group, then the search ends successfully. Otherwise, the function calls itself, passing the new node and the current transistor as parameters.

When the transistors are added to the group, they are removed from the global device list and added either to a list of PMOS transistors or to a list of NMOS transistors, depending on the type of the device. When the grouping is complete, the result is a list of transistor group. Each group contains a separate list of PMOS devices and NMOS devices. The groups will not contain any pass transistors. Pass transistors remain ungrouped and can be found in the main device list.

Once the circuit has been partitioned into groups, the recognition process begins. The algorithm used to recognize and replace complex gates is shown in Algorithm 8.2.

Algorithm 8.2 (The iSPLIT Complex Gate Recognition Algorithm)

```
complexgate(group) {
    type = determineGroupType(group);
    while (group.numN Dev > 1) {
        if (type == CMOS) {
            if (buildSeriesChain(group.nTree) == SUCCESS) {
                if (fndCorrespParallel(group.pTree) == SUCCESS)
                    makeEquivGate(group, CMOS);
                else {
                    noStructureFound = TRUE; break;
                }
            }
            else if (buildSeriesChain(group.pTree) == SUCCESS) {
                if (fndCorrespParallel(group.nTree) == SUCCESS)
                    makeEquivGate(group, CMOS);
                else {
                    noStructureFound = TRUE; break;
                }
            }
            else { noStructureFound = TRUE; break; }
        }
        else if (type == NMOS or type == pseudoNMOS) {
            if (buildSeriesChain(Group.nTree) == SUCCESS)
                makeEquivGate(group, NMOS);
            else if (getParallelTrans(group.nTree) == SUCCESS)
                makeEquivGate(group, NMOS);
            else {
                noStructureFound = TRUE; break;
            }
        }
    }
    if (noStructureFound == TRUE)  restoreCircuit(group);
    else  finishCircuit(group, type);
}
```
□

It combines parallel and series transistors into equivalent transistors until the group has been reduced to an equivalent inverter. This is a standard simplification procedure used in many programs.

First, each group is classified as an NMOS circuit, a CMOS circuit, or a pseudo-NMOS circuit. An NMOS circuit will have one depletion NMOS transistor which has its source and gate connected together. A CMOS circuit has equal numbers of NMOS and PMOS transistors. A pseudo-NMOS circuit has a single PMOS transistor which has its gate connected to $\mathbf{V}_{SS}$. NMOS and pseudo-NMOS circuits can be treated the same way because their n-trees are identical. Any group that does not fit into one of these patterns is not processed any further.

The procedure for a CMOS circuit is depicted in Fig. 8.5. The program searches for a chain of series transistors in the n-tree or the p-tree. If a chain is found, then it searches for a corresponding set of parallel transistors in the opposite tree. The input nodes of all the transistors in the opposite set must correspond to the input nodes of the transistors in the series chain. If a series chain and a corresponding set of parallel transistors are both found, they are each replaced with an equivalent transistor. At the same time a gate is introduced into the circuit representing the logic performed by the transistors that have been replaced as shown in Fig. 8.5. The input capacitance (**ci**) of this gate is calculated at this time from the sum of the gate capacitances of all the transistors that were connected to one of the input nodes. This process is repeated until the group has been reduced to a number of gates and an equivalent inverter. At this time, the program calculates the model parameters for the group from the equivalent inverter. It then removes the equivalent inverter and inverts the output of the last gate as depicted in Fig. 8.5.

For an NMOS or pseudo-NMOS circuit, the algorithm is similar. The program searches for a series chain of transistors. If a chain is found, the

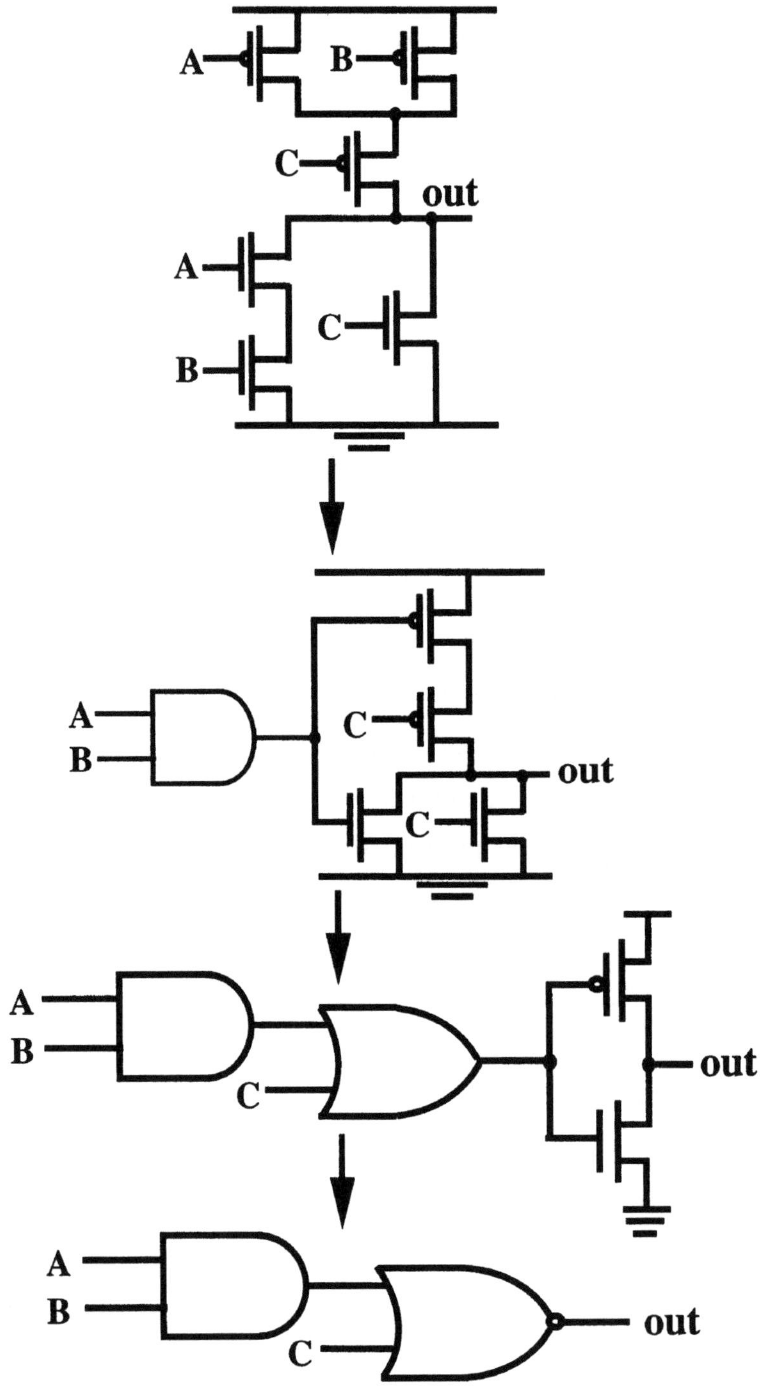

Figure 8.5: Circuit Netlist of a Group During Complex Gate Recognition

program immediately substitutes it with a gate model. Otherwise, the program will look for a set of transistors in parallel. The algorithm continues until the n-tree consists of only one equivalent transistor. For a set of parallel transistors, the parameters of the equivalent transistor are calculated as follows: each device capacitance of the equivalent transistors ($\mathbf{C_{gd}}$, $\mathbf{C_{gs}}$, $\mathbf{C_{bd}}$, and $\mathbf{C_{bs}}$) is calculated by adding the corresponding capacitance value of all the parallel transistors. The beta of the equivalent transistor is taken to be the minimum of the betas of all the transistors in parallel as shown in Fig. 8.6.

For a series chain of transistors, the capacitance values are summed in the same way. However, if one end of the chain is connected to $\mathbf{V_{dd}}$ or $\mathbf{V_{ss}}$, one of the capacitances from the transistor that is connected to the supply is not included. Consider the circuit in Fig. 8.7. If the gate of transistor $\mathbf{M_1}$ is on, and transistor $\mathbf{M_2}$ is initially off and then turns on, it will have to discharge both node X and node Y to ground. The total capacitance used in this case is $\mathbf{C_{gd1}+C_{gs1}+C_{bd1}+C_{bs1}+C_{bd2}+C_{gd2}}$. Therefore $\mathbf{C_{bs2}}$ is removed from further consideration. The beta of the equivalent transistor is given by $\beta = 1/(1/\beta_1+1/\beta_2)$.

The model parameters for the group are calculated after the group has been reduced to an equivalent inverter and placed on the output gate of the group. As described in Section 5.3 and 5.4, there are nine additional parameters that need to be calculated: **co, tr, tf, trc, tfc,** $\mathbf{V_{OH}}$, $\mathbf{V_{OL}}$, $\mathbf{V_{IH}}$, **and** $\mathbf{V_{IL}}$ (the **ci** parameter has already been calculated for all the gates in the group at this point). The gate output capacitance, **co**, is calculated by summing all the capacitances connected to the output node. In addition, any capacitances connected to the internal nodes in a group are added to the total output capacitance, since, in the worst case, they must be charged when the gate switches. Next, the four noise margin parameters $\mathbf{V_{OH}}$, $\mathbf{V_{OL}}$, $\mathbf{V_{IH}}$ and $\mathbf{V_{IL}}$ are computed. For CMOS circuits, all these parameters

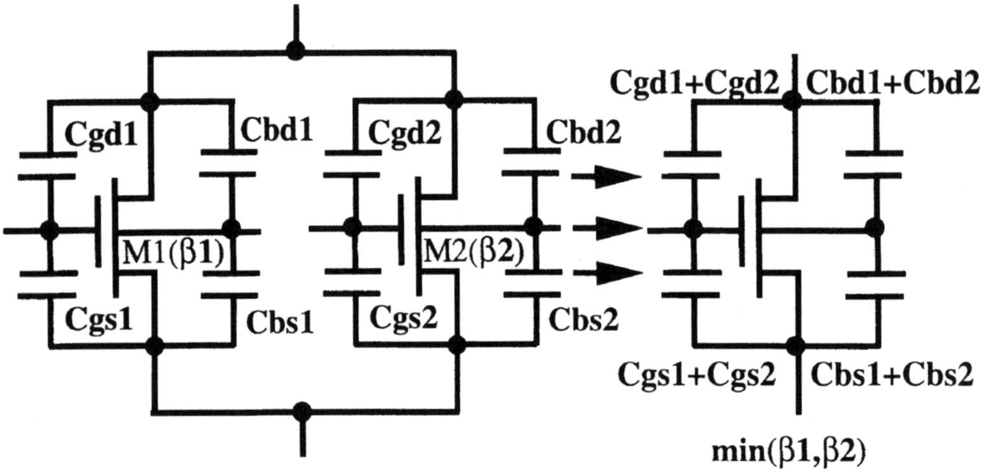

Figure 8.6: Equivalent Transistor for Two Parallel Transistors

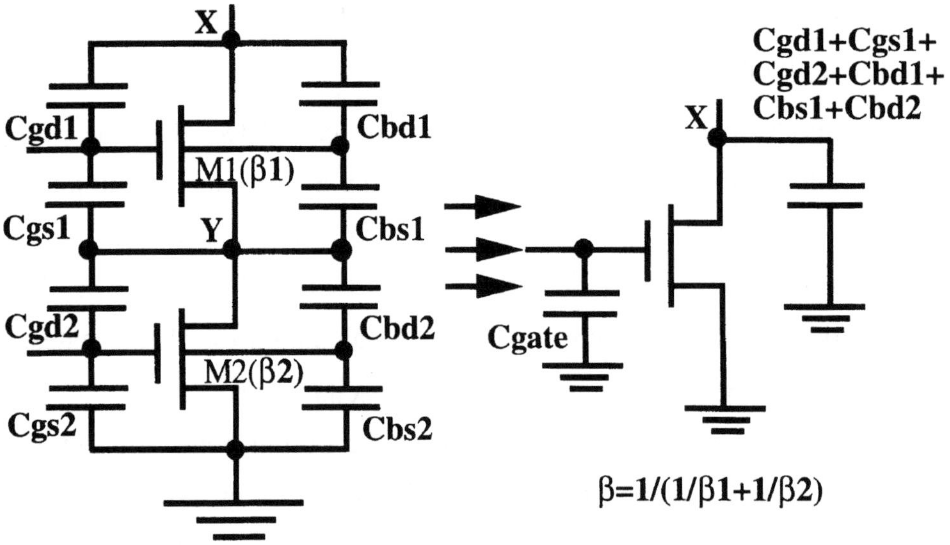

Figure 8.7: Equivalent Transistor for Two Series Transistors

are calculated using analytic formulas. The parameter $\mathbf{V_{OH}}$ is simply equal to $\mathbf{V_{dd}}$, and $\mathbf{V_{OL}}$ is equal to $\mathbf{V_{ss}}$. The input parameters, $\mathbf{V_{IH}}$ and $\mathbf{V_{IL}}$ are calculated by formulas provided in [WES85]. For NMOS circuits, $\mathbf{V_{OH}}$ is simply equal to $\mathbf{V_{dd}}$. The other parameters are calculated by iterative equations in [UYE88]. Finally, the parameters related to the rise and fall times of the equivalent inverter are calculated using the equations provided in Section 5.4.

8.5.3. Recognition of User-Defined Components

The techniques described above are useful for MOS logic circuits but are not suitable for bipolar transistors, analog circuits or high-level block recognition. For example, an analog comparator or an operational amplifier may contain 30-50 transistors and the connections and device sizes may be different for each design. Clearly, this problem is much more difficult than recognizing complex digital gates since the channel-connected groups could be examined individually and were known to have a particular structure. With user-defined components, however, the structure to be identified may contain two or more groups. This is also the case when recognizing higher-level blocks such as flip-flops from lower level primitives such as NAND gates.

To facilitate the recognition and replacement of these types of blocks, the user defines components in a library file that iSPLIT reads during startup. After the startup file has been read, the iSPLIT program forms two netlists for every user-defined component: a netlist for the cell, and a netlist for the higher-level model. The transistor netlist is broken up into channel-connected groups in the same manner as the main circuit.

Once all groups in the user cells have been processed, iSPLIT regroups all groups that have not been replaced with logic models (both in the main circuit and in all the user-defined components). The regrouping proceeds in the same manner as the initial grouping, except that there is no restriction on the number of possible output nodes. The next step in the process is to examine all groups in all user-defined components that are still at the electrical level, and assign each one a unique type number, or *signature*, that depends on the circuit topology of the group. The iSPLIT program builds a table which contains entries for every unique group. Each entry has its own signature number. Every new group is compared against all the existing group using their signatures. If it matches one of them, it is assigned the signature number corresponding to that group. Otherwise, the group is assigned a new signature which is added to the table.

The signature calculation is based on an algorithm published by Beatty and Bryant [BEA88]. This algorithm was originally used to speed up the preprocessing of a switch-level simulator. The algorithm works by assigning each group a unique signature that depends on the circuit topology. This number is then stored in a hash table. Any new group can be instantly compared against all existing group types by looking for its hash number in the table. If the number exists, the group can be assigned the type number that corresponds to that hash number.

Once all groups in the user-defined cells have been processed, all electrical-level groups in the main circuit are processed. Each group is compared against all group types in the table. If it matches one of the entries, it is given the corresponding type number. In addition, it is added to a list of instances of that type of group. Otherwise, it is ignored.

Recognizing user-defined cells belongs to the generalized problem of subgraph isomorphism. Many heuristics have been published to solve this problem, but not all of them are applicable to the problem being addressed here. In iSPLIT, a very simple tree-based algorithm is used. The program starts with one element from a user-defined cell, and an element of the same type from the main circuit. Then, all the corresponding pins on each of the two elements are checked to see if they have they same number and type of elements connected to them. Next, all elements connected to those pins are recursively checked in the same manner. The search proceeds until the program reaches an I/O pin on the user cell or there is a mismatch.

The iSPLIT program begins looking for a specific user cell by finding the primitive in the cell with the highest type number. Since the lower type numbers are standard elements such as logic gates, and the higher type numbers are groups with computed type numbers, there should be fewer instances of these elements. For example, the type number 52 may represent a group that appears eight times in the circuit, while the type number 12 may represent an inverter that appears in the circuit a few hundred times. Choosing the largest type number in this case narrows the search space considerably. However, in some circuits, such as D-flip-flops, the highest component may be a four-input NAND gate. In this case, the algorithm will not be very efficient, because the program must investigate every four-input NAND gate and there may be thousands of these in the circuit.

Run times for iSPLIT are very fast. For circuits of a few hundred transistors, the runtime is a few seconds or less. For larger circuits, the runtime may be on the order of minutes, and will be longer if the user has defined cells for analog macromodels. However, compared to the time required for simulating the circuit at the electrical level, the runtimes are quite acceptable.

8.6. MIXED-MODE SIMULATION EXAMPLE

In this section, a 1K-bit CMOS static RAM, which was implemented based on an existing industrial design as shown in Fig. 8.8, is used as an example for mixed-mode simulation [SAL94B]. The focus here is mainly on the overall speed of simulation while producing the correct results. The speed comparisons are representative of typical mixed-mode simulators available today.

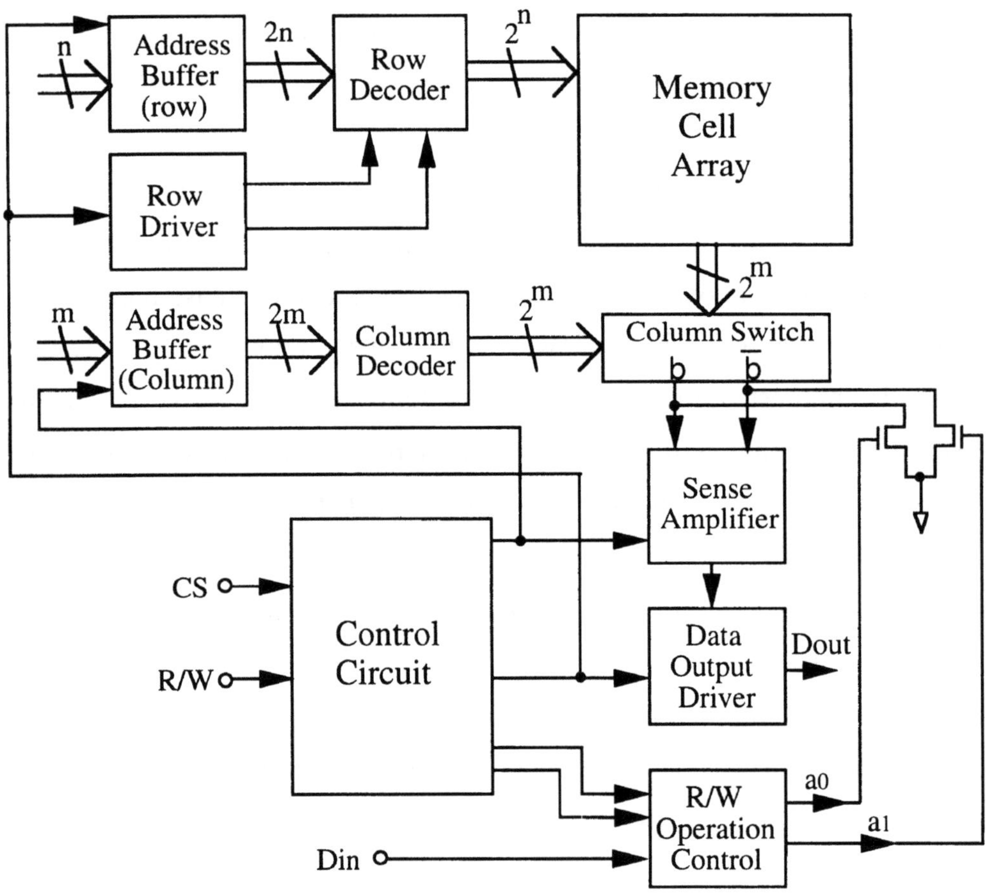

Figure 8.8: 1K-bit Static RAM

Memory circuits are excellent applications for mixed-mode simulators since they combine digital blocks such as decoders, buffers, and flip-flops with analog blocks such as lumped-RC networks, pass transistors and sense amplifiers. Furthermore, the size of these circuits can often exceed the limitations of SPICE. In the architecture of this particular memory circuit, the memory cells are organized in a 32x32 configuration. Each cell consists of 6 transistors. Hence, 10 address lines are used in the row and column decoders to select one of the 1024 cells available. A sense amplifier is used to amplify the voltage difference between the bit and bit bar lines and the output is sent to a tri-state data output driver. This 1K-bit RAM contains 7698 transistors, 128 resistors, 208 external capacitances and has a total of 2908 nodes.

For mixed-mode simulation, all the peripheral circuits can be simulated in the digital domain while the memory cell array, column switches and sense amplifiers should be simulated in the analog domain. To reduce the simulation time, the local feedback cross-coupled inverters in each memory cell should be modeled in digital domain, as depicted earlier in Fig. 7.9. Manually converting the circuit into a mixed-mode description is very time-consuming and sometimes leads to inaccuracies, as explained above. This is because the logic gates must first be recognized and then the parameters of the gates must be extracted precisely so that correct results can be obtained when the digital-to-analog interface is encountered. Therefore, an automatic mixed-mode partitioner, such as iSPLIT, should be used to speed up the conversion process. After iSPLIT is employed, the circuit contains 2293 gates at logic level and 2472 transistors at the electrical level (mixed-mode II). iSPLIT required only 10 seconds to perform the conversion. If the pair of cross-coupled inverters in each memory cell are kept at the electrical level, then the circuit contains 6568 transistors at the electrical level (mixed-mode I). Table 8.1 lists the simulation results of 1K-bit static RAM.

As the results indicate, mixed-mode simulation in iSPLICE3 is over 40 times faster than a traditional circuit simulator such as PSPICE. Part of the speed improvement is provided by the block ITA method which is 5 to 10 times faster than traditional circuit simulation. The remainder of the speedup, of roughly 5, is provided by mixed-mode simulation. In general, the speedup for a given circuit will depend on the number of transistors that are simulated at the electrical level since this is the most expensive mode of simulation. This is obvious when the runtimes of mixed-mode I and mixed-mode II are compared. In the final mixed-mode circuit, approximately one-third of the circuit is still at the transistor level, so the speedup is not expected to be very large. Note also that, since most of the cells are not selected during the operation of the memory circuit, they

Program	Transistor			CPU-time (sec.)	Speedup
	Electrical	Logic	% of Analog		
PSPICE version 6.0	7698	0	100	35125	1
iSPLICE3 electrical	7698	0	100	4099	8.6
iSPLICE3 mixed-mode I	6568	1130	85.3	1935	18.2
iSPLICE3 mixed-mode II	2472	5226	32.1	822	42.7

Table 8.1: Performance Comparisons on SUN SPARCstation 10.

remain inactive over long periods of time. In fact, the activity of very large memory circuits is extremely low since only one cell is selected in any given cycle (although all the cells in a row may be turned on during the operation). This limits the speed improvement when the cross-coupled electrical inverters are replaced by logic inverters. However, the speedup over direct methods will continue to grow with the memory circuit size due to the cost of the matrix solution.

To illustrate the accuracy aspects of mixed-mode simulation, Fig.8.9 shows the waveforms obtained by using pure electrical level simulation and the mixed-mode II partitioning described above. In Fig. 8.9, two critical nodes are compared. One is the output of the pre-sense amplifier and the other is the data output bit (DOUT). As the figure indicates, the results of the two simulations are very similar except at the beginning of the transient solution of the pre-sense amplifier output. This is because there is a small timing difference in the bit and bit bar lines in the two simulations. After the pre-sense amplifier, the difference is amplified and results in a different transient starting point in the output of pre-sense amplifier. Thus, this node is very sensitive in the two types of simulations. However, the output node (DOUT) is almost indistinguishable in the two simulations. Therefore, with proper attention to logic modeling, parameter extraction and the mixed-mode interface, mixed-mode simulation can provide accurate simulation results with substantially shorter runtimes compared to traditional circuit simulation.

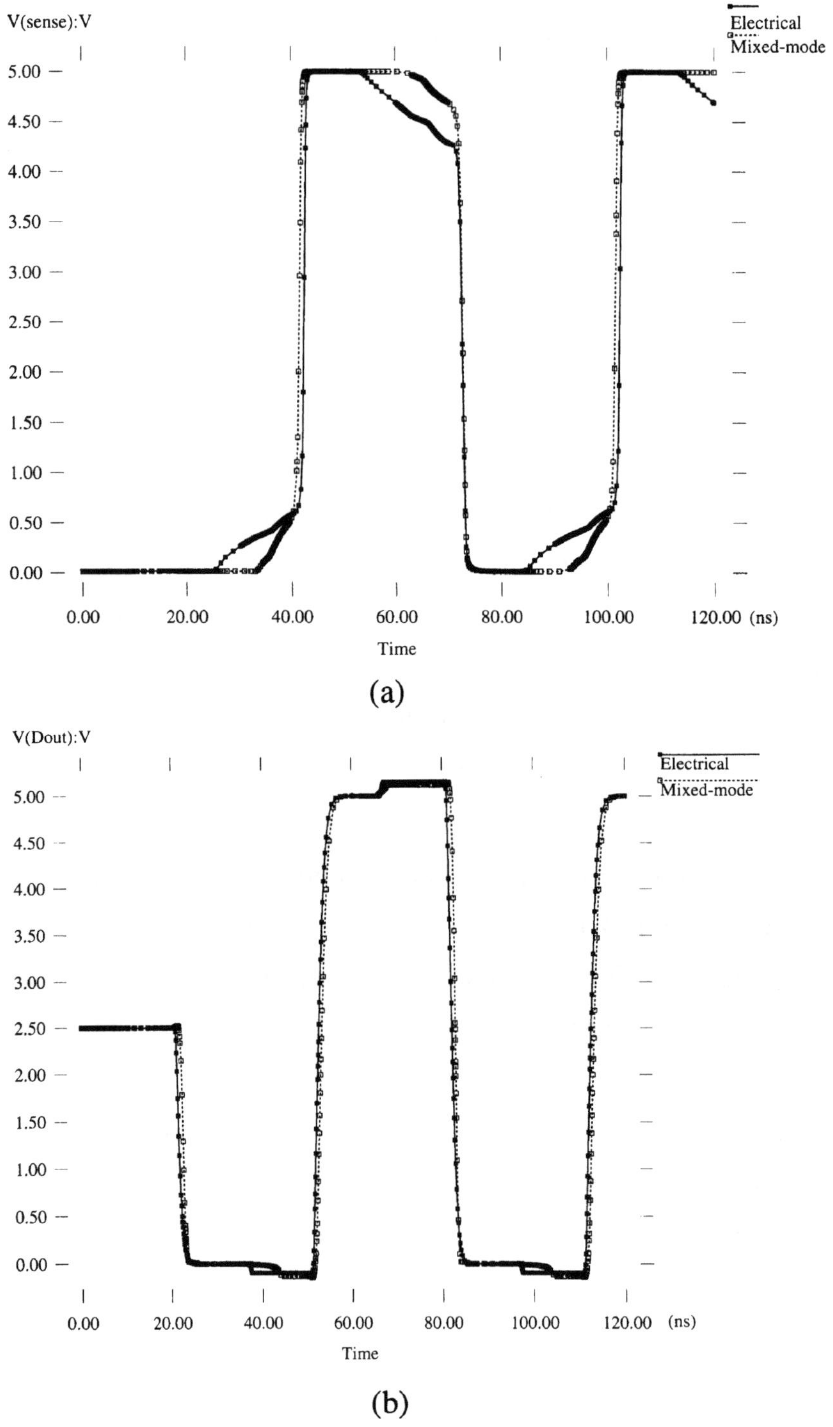

Figure 8.9: Simulation Results of 1K-bit Static RAM

CHAPTER 9

ANALOG MULTILEVEL SIMULATION

9.1. INTRODUCTION

SPICE and its derivative programs remain the primary simulation tools in use today by analog designers. However, over the past decade analog circuit designs have increased in complexity to the point where the basic techniques used in SPICE are not fast enough to produce a solution in a reasonable amount of time. Currently, system-level design, modeling and simulation are being emphasized to cope with the complexities of these large designs. As in the digital case, an analog designer would like to specify portions of the analog circuit at a higher level of abstraction in order to carry out functional verification. For example, the designer may wish to represent a filter block in terms of an s-domain transfer function, rather than specifying all the transistors and their interconnections, to evaluate a proposed architecture of an analog system before a detailed design begins. And ideally, the entire system could be specified in some form of standard analog hardware description language (AHDL) similar to the languages that have emerged for digital hardware description.

An example of a typical analog system is shown in Fig. 9.1 which depicts an oversampled A/D converter. This circuit consists of a filter block followed by a sample-and-hold stage, a $\Sigma\Delta$ modulator and a decimation filter. The circuit is driven by clocks which have high frequencies relative to the input signals of interest. In this type of circuit, the designer would like to explore the design space and evaluate certain design tradeoffs at the system level. In order to accomplish this, the high-level models and parameters for each block could be specified in some form of AHDL,

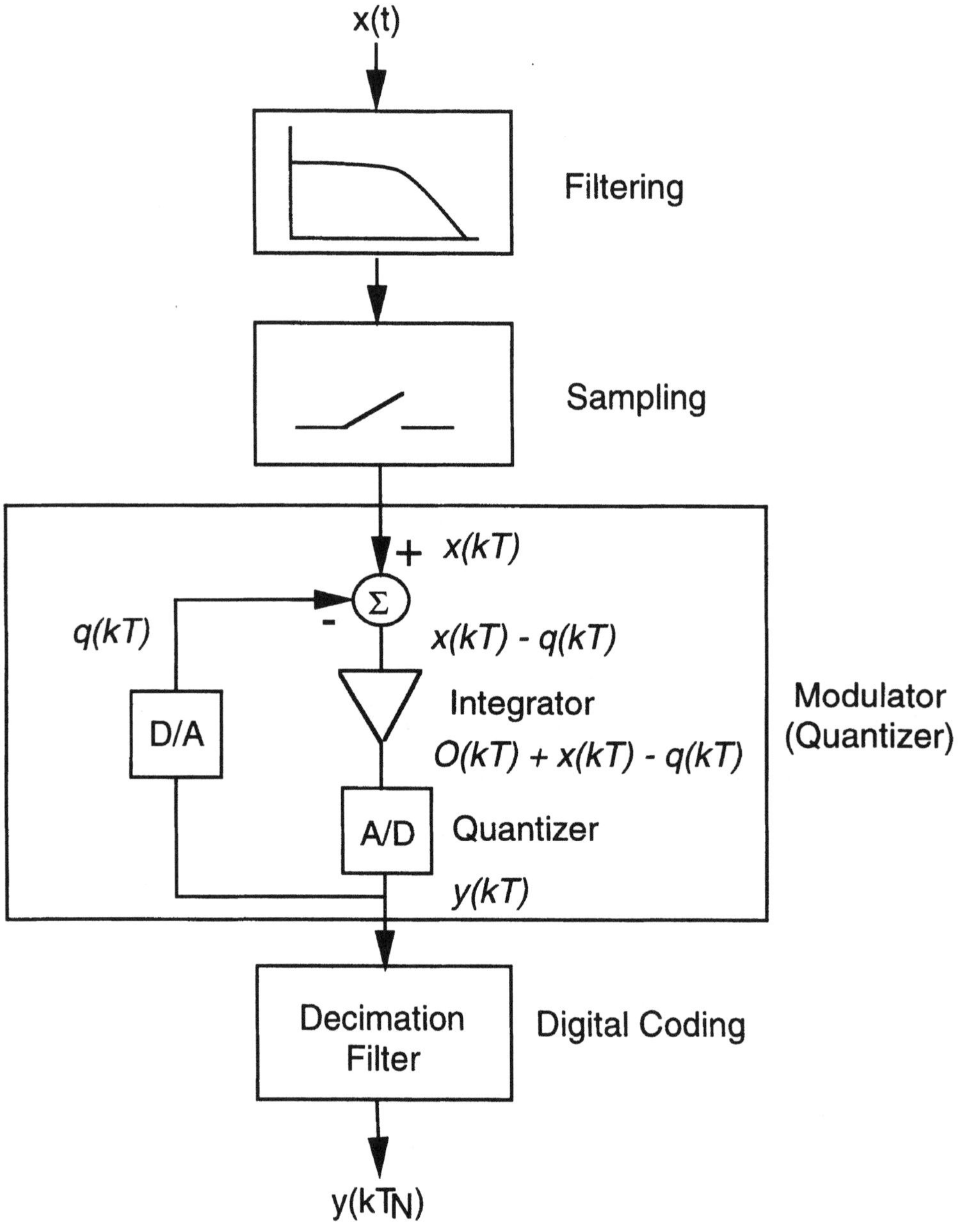

Figure 9.1: Block Diagram of Oversampled A/D Converter

and the analysis could be performed using a high-level simulator to determine if the overall architecture is suitable. If so, each block could be individually designed to meet the desired specifications.

On the other hand, if the entire circuit were simulated at the electrical level, the time-steps used in the simulation would be governed by the highest frequency clock signal driving the circuit and the smallest time constants in the circuit. Therefore, a sizable number of time points would be computed for each clock period. Since the simulation interval of interest to the designer in these circuits usually involves a large number of clock cycles, the overall simulation would be prohibitively expensive. So, to simulate systems of this type, new tools are needed that are well beyond the capabilities of SPICE.

To address this complexity issue, an important goal in analog simulation has been to develop a multilevel simulation environment that incorporates all of the different levels of simulations shown in Fig. 1.1, and described in Sections 1.3 and 1.4. Great progress has been made in this area recently and a number of simulators have been developed to fill the need. These includes AMP [RUM89], ATTSIM [ATT91], DESIGN CENTER [MIC94], ELDO [ANA93], iMACSIM [SIN91], ARCHSIM [ANT93], CONTEC SPICE [CON93], MIDAS [WIL92], $\mathbf{M}^3$ [CHA92] and SABER [VLA90], to name a few. Table 9.1 contains a partial list of commercial analog multilevel simulator. iMACSIM [SIN91] is included as a reference for some of the implementation mechanisms described in this book. Note that because there is not yet a mature and standard analog HDL, the language used for behavioral description is quite different among the simulators. Nevertheless, **s**-domain/**z**-domain transfer function, differential equation and difference equation are commonly provided to describe the behavior of analog system.

Program/Company	*Type*	*Behavioral*	*Ideal/Non-Ideal*	*Electrical*
AccuSim II Mentor Graphics Corp.	I	HDL-A, S, Z, DT	A/D, D/A, Switch, Sample/Hold, Peak Detector etc.	Direct Matrix
ATTSIM AT&T-Design Automation	I	C Language, S, Z, DT, DS	Same as Behavioral	Direct Matrix
ContecSPICE Contec Inc.	I	C primitives, S, Z, DT	Switch, Integrator, Arbitrary Func. etc.	Direct Matrix
Eldo Anacad Inc.	I	HDL-A, S, Z, DT, DS	Integrator, Opamp, S/H, PWM, VCO, etc.	ITA and Direct Matrix
PSPICE Microsim Corp.	I	ABM , S, Expression Func.	Math. Func., Filter, Limiter, Table Look-Ups etc.	Direct Matrix
Saber Analog Inc.	I	MAST, S, Z, DT, DS	VCO, PLL, A/D, D/A, opamps, etc.	Direct Matrix
Spectre Cadence Inc.	I	Spectre-HDL, S, Z, Profile, DT, DS	VCO, PLL, A/D, D/A, Mux, opamps, etc.	Direct Matrix
iMACSIM University of Illinois	I	Analog HDL, S, Z, DT, DS	Switch, Control sources	ITA and Direct Matrix

Legend S: S-domain transfer function, Z: Z-domain transfer function, DT: Differential equation, DS: Difference equation

Table 9.1: Survey of Analog Multilevel Simulators and Their Capabilities

In this chapter, techniques for the simulation of analog systems in time-domain are described. Issues in analog multilevel simulation are addressed in Section 9.2. Behavioral level simulation methodologies of continuous-time and discrete-time systems are discussed in Sections 9.3 and 9.4. Mixed continuous/discrete simulation is described in Section 9.5. The architecture of the iMACSIM multilevel analog circuit simulator is outlined in Section 9.6. Then, two analog multilevel simulation examples are illustrated in Section 9.7. Finally, in Section 9.8, a macromodeling and simulation environment are described. Although frequency-domain simulation is also very important for analog circuits, it is beyond the scope of this book. The interested reader may refer to [KUN90] for details on this subject.

9.2. SIMULATION ISSUES

Many systems today are designed using a combination of discrete-time components and continuous-time components. For example, a commercial stereo codec chip uses a continuous-time anti-aliasing filter at the input, a discrete-time ΣΔ A/D converter implemented in switched-capacitor technology, and a number of digital signal processing (DSP) blocks. The DSP blocks are followed by a D/A converter and another analog filter block which selects the signal for eventual output. A complete simulation of this chip in a program such as SPICE would not be feasible. In fact, SPICE does not have a true discrete-time simulation capability for switched-capacitor simulation. However, a system of this type could be efficiently simulated using a mixture of continuous-time methods of SPICE and the discrete-time methods of a program such as SWITCAP [FAN83]. This issue of mixed continuous/discrete simulation is discussed in Section 9.5.

The simulation problems posed by analog designs are not solely a function of their larger sizes. Many small circuits, such as phase-locked loops (PLL) and switched-capacitor filters, can be very time-consuming to simulate if a SPICE-like program is used. This is due primarily to the disparity between the relatively large simulation intervals of interest to the analog designer and the small time-steps that SPICE takes during the solution process [MA92]. For example, when simulating a PLL, one is interested in the amount of time it takes for the circuit to lock on to the input frequency. The locking process may take hundreds or even thousands of clock cycles to occur. This often translates into days of actual simulation time. To be able to simulate these circuits in a reasonable amount of time without sacrificing needed accuracy, the designer needs an event-driven multilevel simulation environment which supports macromodeling and the optimization of models.

Analog behavioral modeling and simulation are still open areas of research [RUT93, SAL94A, SIN94A]. A standard high-level description language is still needed for analog applications ranging from low frequencies up to microwave frequencies. The behavior of a system should be expressible in the time-domain or frequency-domain. In the time-domain, continuous systems can be mathematically characterized by linear or nonlinear differential equations, and discrete systems by difference equations. In the frequency-domain, the corresponding representations are algebraic expressions in terms of the Laplace transform variable, **s**, and the **z**-transform variable, **z**. Since the behavior can be specified in either the time-domain or the frequency-domain, it is very important that efficient algorithms be used to perform transformations between these different model representations. The techniques to simulate analog circuits behaviorally are outlined in Sections 9.3 and 9.4.

9.3. CONTINUOUS-TIME BEHAVIORAL MODELS

At the behavioral level it should be possible to describe individual blocks in terms of **s**-domain transfer functions or using differential equations, and models written in AHDL. Each of these options is described below.

9.3.1. Behavioral Models using a Hardware Description Language

It is widely-accepted that a standard analog hardware description language (AHDL) is needed for the description, documentation and exchange of analog design data. An AHDL would allow the designer to quickly define the structure and behavior of new blocks needed for a simulation. Some simulators have user-defined controlled sources that allow a variety of different models to be incorporated into the simulator. However, there are limitations on the type of elements that this approach can handle. Other simulators currently allow this to be performed using an existing programming language. A C or C++ file is created containing the functions needed by the rest of the simulator. The designer specifies the input and output nodes, partitioning information, setup and preprocessing information, and scheduling information. These functions, along with the necessary data structures, are then compiled and linked to the program to create an executable routine that includes the new behavioral block. Of course, analog designers are not eager to use embrace this approach because it is not a natural way to specify an analog design.

In other cases, proprietary languages have been developed for this purpose. Each language is different from the other and this does not allow models to be exchanged between programs easily. A standard AHDL would greatly enhance the ability to perform behavioral simulation because a large database of analog components could be developed and reused by different customers and vendors. Currently, MHDL [MHL91] and VHDL-A [VHD91] are the two languages under development with the intention of eventually being standardized. While space does not permit us to provide the details of these efforts, the reader is encouraged to survey the literature on this important activity.

9.3.2. s-domain models

The general behavior of a system can be viewed as a transformation of a set of input signals into a set of output signals. Thus, a suitable representation for the analog behavior of a system is the input/output transfer function. For a linear, time-invariant, continuous-time system, the s-domain representation of the transfer function is usually specified as **H(s)**. This is a frequency-domain characterization of the impulse response of a linear system, **h(t)**. It provides valuable information to the analog designer about the magnitude and phase response of the system and is, therefore, preferable to the time-domain representation. On the other hand, a time-domain representation can capture all the nonlinearities of the system and should be used if these nonlinearities are important to the overall performance.

The transfer function, **H(s)**, provides information about the system poles and zeros that indicate the stability of the system. It is normally expressed as a ratio of polynomials, as follows:

$$\mathbf{H(s)} = \frac{\mathbf{a_0 + a_1 s + \cdots + a_p s^p}}{\mathbf{b_0 + b_1 s + \cdots + b_{q-1} s^{q-1} + s^q}} + \mathbf{r} = \frac{\mathbf{N(s)}}{\mathbf{D(s)}} + \mathbf{r} \qquad (9.1)$$

where $\mathbf{p < q}$ and $\mathbf{r}$ is a remainder term. For simulation, **H(s)** must be first transformed into the time-domain. Several methods have been proposed to perform this transformation as described below. The techniques are categorized according to their modeling and the mathematical methods used.

1) State-space representation and numerical integration method

This approach is perhaps the most popular of all the methods to be described [CHA92, VIS88, SIN91, TRI90]. For an input excitation U(s), the output Y(s) is given by:

$$\mathbf{Y(s) = H(s)U(s) = \frac{N(s)}{D(s)}U(s) + rU(s)} \tag{9.2}$$

Define an auxiliary system M(s), given by

$$\mathbf{M(s) = \frac{U(s)}{D(s)}} \tag{9.3}$$

and Eq. (9.3) can be rewritten as

$$\mathbf{(s^q + b_{q-1}s^{q-1} + \cdots + b_1 s + b_0)M(s) = U(s)} \tag{9.4}$$

The corresponding **q**-th order differential equation is:

$$\mathbf{m^{(q)}(t) + b_{q-1}m^{(q-1)}(t) + \cdots + b_1 m^{(1)}(t) + b_0 m(t) = u(t)} \tag{9.5}$$

We now define **q** state variables as follows:

$$\mathbf{x_1(t) = m(t)}$$

$$\mathbf{x_2(t) = \dot{x}_1(t)}$$

$$\vdots$$

$$\vdots$$

$$\mathbf{x_q(t) = \dot{x}_{q-1}(t)}$$

Then, it follows that:

$$\mathbf{x_2(t) = m^{(1)}(t)}$$

$$\mathbf{x_3(t) = m^{(2)}(t)}$$

$$\vdots$$

$$\vdots$$

$$\mathbf{x_q(t) = m^{(q-1)}(t)}$$

Substituting the above equations into Eq. (9.5) we obtain:

$$\dot{\mathbf{x}}_q = -\mathbf{b}_{q-1}\mathbf{x}_q - \mathbf{b}_{q-2}\mathbf{x}_{q-1} - \cdots - \mathbf{b}_0\mathbf{x}_1 + \mathbf{u(t)} \tag{9.6}$$

and the output can be expressed as:

$$\mathbf{Y(s) = (a_0 + a_1 s + \cdots + a_p s^p)M(s) + rU(s)} \tag{9.7}$$

In the time-domain, with state variables substituted for the derivatives of **m** we obtain:

$$\mathbf{y(t) = a_0 x_1(t) + a_1 x_2(t) + \cdots + a_p x_{p+1}(t) + ru(t)} \tag{9.8}$$

Therefore, the controllable canonical-form realization of this transfer function is:

$$\dot{\mathbf{x}} = \mathbf{Ax + Bu} \tag{9.9a}$$

$$\mathbf{y = Cx + Du}$$

where

$$\mathbf{A} = \begin{bmatrix} 0 & 1 & 0 & .. & 0 \\ 0 & 0 & 1 & .. & 0 \\ 0 & 0 & 0 & .. & 0 \\ \vdots & \vdots & \vdots & \vdots & \vdots \\ 0 & 0 & 0 & .. & 1 \\ -\mathbf{b}_0 & -\mathbf{b}_1 & -\mathbf{b}_2 & .. & -\mathbf{b}_{q-1} \end{bmatrix} \quad \mathbf{x} = \begin{bmatrix} \mathbf{x}_1 \\ \mathbf{x}_2 \\ \mathbf{x}_3 \\ \vdots \\ \mathbf{x}_{q-1} \\ \mathbf{x}_q \end{bmatrix} \quad \mathbf{B} = \begin{bmatrix} 0 \\ 0 \\ 0 \\ \vdots \\ 0 \\ 1 \end{bmatrix}$$

$$\mathbf{C} = [\mathbf{a}_0 \quad \mathbf{a}_1 \quad \mathbf{a}_2 \quad \cdots \quad \mathbf{a}_p \quad 0 \quad \cdots \quad 0] \quad \mathbf{D = r}$$

and **x** is the vector of state variables. The system can also be represented in the observable form which is essentially the transpose of the controllability form [CHE84]. That is, if $\mathbf{A_o}$, $\mathbf{B_o}$, $\mathbf{C_o}$ and $\mathbf{D_o}$ are the system matrices of a corresponding observability form, then the two forms are related as follows:

$$\mathbf{A_o} = \mathbf{A^T};\ \mathbf{B_o} = \mathbf{C^T}; \qquad (9.9b)$$

$$\mathbf{C_o} = \mathbf{B^T};\ \mathbf{D_o} = \mathbf{D}$$

where $\mathbf{A^T}$ refers to the matrix transpose of A.

With the above transformations, the result is a system of first-order linear differential equations that can be solved using numerical integration, as described in Chapter 2. Each unidirectional **s**-domain block can be solved separately with its own time-step control to keep local truncation error bounded within specified limits. The time-step control algorithm discussed in previous chapters for electrical-level subcircuits can also be applied here. At a given time point the algorithm will compute the next acceptable time-step for each **s**-domain block, which will be scheduled accordingly. It is important to note that it is not possible to determine the initial output values of a transfer function block for a time-domain transient analysis by setting **s**=0 in the transfer function. This limit specifies an initial value for steady-state analysis. Therefore, for a transient analysis the state variables must be initially be set to zero, and the initial output value must be user-specified to maintain consistency with the rest of the circuit.

In addition to the unidirectional **s**-domain functions discussed above, there are cases in which an **s**-domain transfer function may have both its input variable and its output variable connected to the same circuit node. For example, the input variable to the block could be current and the output variable could be voltage, or vice-versa. To process such "bidirectional" connections, the state-variable equations of the block must be embedded in an MNA formulation of the circuit connected to it. Then, the **s**-domain state equations can be solved simultaneously with the circuit-level equations.

2) State-space Representation and Power Series Method

A less popular alternative [TAH89] to the above approach is to convert the solution of Eq. (9.9a) into a set of difference equations, as described in [CHU75]. First, the state-space equation is formulated as shown above. The solution for $\mathbf{x(t)}$ in Eq. (9.9a) is given by [CHU75]:

$$\mathbf{x(t)} = \mathbf{e^{At}}\int_{t_0}^{t}\mathbf{e^{-A\tau}Bu(\tau)d\tau} + \mathbf{e^{A(t-t_0)}x(t_0)} \tag{9.10}$$

where $\mathbf{e^{At}}$ is defined by the following infinite power series:

$$\mathbf{e^{At}} = 1 + \mathbf{At} + \frac{1}{2!}(\mathbf{At})^2 + \cdots + \frac{1}{\mathbf{n}!}(\mathbf{At})^{\mathbf{n}} + \cdots. \tag{9.11}$$

These equations are substituted into $\mathbf{y(t)}$ in Eq. (9.8) to obtain the overall solution.

Many different approaches to numerically compute the solution to $\mathbf{x(t)}$ exist. One possible discretization of Eq. (9.10) is as follows [CHU75]:

$$\mathbf{x[(k+1)T]} = \mathbf{e^{AT}x(kT)} + \mathbf{e^{AT}B}\frac{\mathbf{T}}{2}\mathbf{u(T)} + \mathbf{B}\frac{\mathbf{T}}{2}\mathbf{u[(k+1)T]}$$

where T is the time unit. This equation can be used to sequence through the solution at different time points during the simulation. The main problem with this approach is that the matrix exponential, $\mathbf{e^{AT}}$, is known to be difficult to compute [MOL78].

3) SPICE-based Macromodels

Another approach to the realization of a behavioral model is to implement a desired $\mathbf{H(s)}$ using basic primitives available in a circuit level simulator such as SPICE. In this method, the first step is to modify $\mathbf{H(s)}$

by dividing each term of the numerator and denominator by the highest power of s:

$$\mathbf{H_m(s)} = \frac{\mathbf{a_0 s^{-q} + a_1 s^{-q+1} + \cdots + a_p s^{p-q}}}{\mathbf{b_0 s^{-q} + b_1 s^{-q+1} + \cdots + b_{q-1} s^{-1} + 1}} + \mathbf{r}$$

Then, Mason's gain formula [CHE84] is used to construct a function block diagram for this system. In this function block diagram, each term of the denominator, except for the constant, represents a feedback path whose coefficient is the feedback gain and each term of the numerator is a feed forward path and its coefficient is the forward gain. As an example, consider the third-order system given by:

$$\mathbf{H(s)} = \frac{\mathbf{a_0}}{\mathbf{b_0 + b_1 s^1 + b_2 s^2 + s^3}}$$

This transfer function is modified by dividing the numerator and denominator by $\mathbf{s^3}$ to produce:

$$\mathbf{H_m(s)} = \frac{\mathbf{a_0 s^{-3}}}{\mathbf{b_0 s^{-3} + b_1 s^{-2} + b_2 s^{-1} + 1}}$$

The corresponding functional block diagram is shown in Fig. 9.2. Based on this block diagram, an equivalent circuit macromodel is constructed using SPICE primitives [CHO89]. The summing point can be realized using a nonlinear polynomial dependent VCVS. For the example above, the output of the summer would have the equation:

$$\mathbf{V_2 = a_0 V_{in} + b_2 V_3 + b_1 V_4 + b_0 V_{out}}$$

For the needed integrator blocks, an approximate circuit model is implemented using a linear VCCS, a resistor and a capacitor as shown in Fig. 9.3. The transfer function of this circuit is:

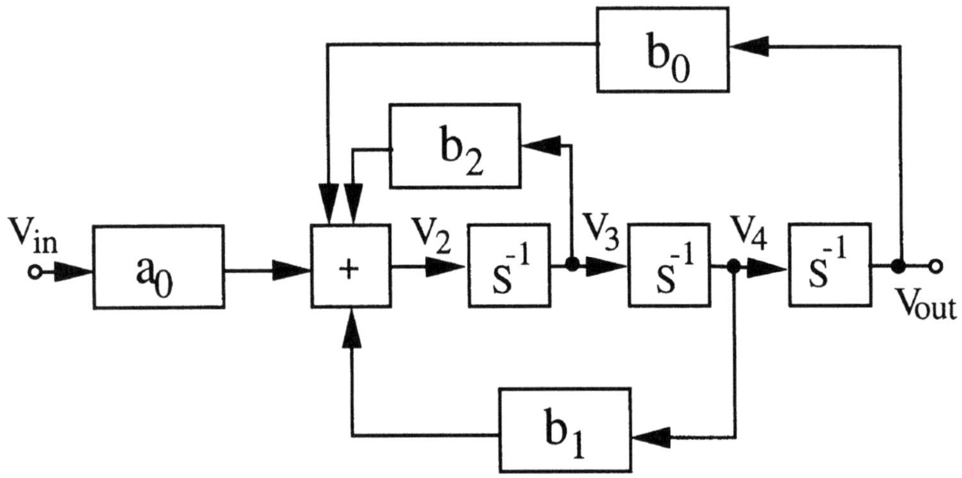

Figure 9.2: Functional Block of $\mathbf{H_m(s)}$

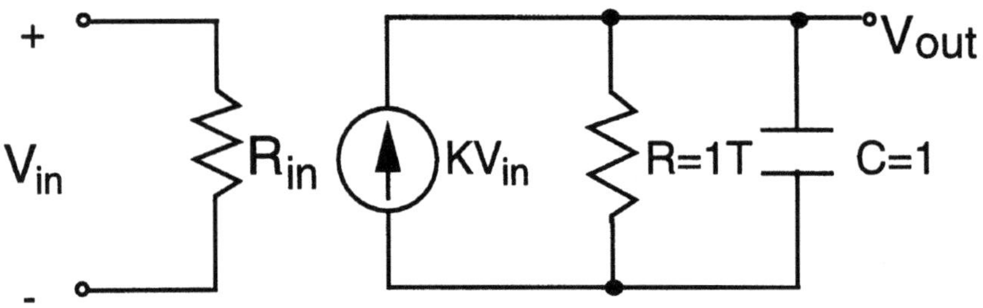

Figure 9.3: An Approximate Circuit Model for An Integrator

$$\mathbf{H_I(s)} = \frac{\mathbf{V_{out}(s)}}{\mathbf{V_{in}(s)}} = \frac{\mathbf{K(1/C)}}{\mathbf{s+(1/RC)}}$$

where K is the gain of the VCVS. If the capacitor value is 1F and the resistor value approaches infinity, the transfer function is:

$$\mathbf{H_I(s)} = \frac{\mathbf{V_{out}(s)}}{\mathbf{V_{in}(s)}} = \frac{\mathbf{K}}{\mathbf{s}}$$

which provides the integration function.

4) Other methods

If the input signal can be expressed in the **s**-domain, then Y(s) may be calculated by Y(s) = H(s)U(s), and the inverse Laplace transform can be applied to obtain y(t). To use this method in the case where a limited number of points of the input signal waveform are available, a Fast Fourier Transform (FFT) [RUM89] can be used to transform u(nT) into U(s). To calculate the inverse Laplace transform, the partial fraction expansion method and numerical Laplace transform inversion (NLTI) are used. In the partial fraction expansion methods, the poles and residues of the function need to be calculated. However, this process is known to be very expensive if **q** is large [HAL88]. Another alternative is to numerically compute the inverse Laplace transform directly. However, it is only suitable for a nonperiod excitation and the error grows as time increases. A modified version of NLTI called the stepping algorithm has been proposed to bound the error but the computation complexity is the same as solving Eq.(9.9) using numerical integration method [VLA83].

Another approach is to take the input waveform and convolve it with the impulse response of the **s**-domain block to obtain the time-domain output. The time-domain impulse response can be obtained from the **s**-domain transfer function through an inverse Fourier integral. However, the convolution approach has been demonstrated to be inferior to the state-variable approach [TRI90]. Its computation time has been shown to increase superlinearly as the simulation interval is increased, assuming that the sampling time is kept constant, and the memory requirements of this approach are very large.

To summarize the above mentioned methods, the state variable approach is preferable because it can be easily embedded into the existing electrical simulation scheme, and uses the same numerical integration method and time-step control. As for the inverse Laplace method, one disadvantage is that the whole input waveform must be known before the s-domain blocks are simulated. Also, computing the Inverse Laplace transform is computationally intensive. The SPICE macromodeling method to represent **H(s)** is most suitable for extending the abilities of SPICE-like simulators.

9.3.3. Differential Equations

The simulator should allow the behavior of a block to be specified in terms of linear or nonlinear differential equations. The simulation algorithm must provide time-step control during the equation solution process. It may not be possible to directly apply the differential equation techniques used in circuit simulation to behavioral blocks, since the circuit-level equations are assumed to be first-order equations. If the given behavioral differential equations are of greater order, they will have to be reduced to a set of first-order equations or solved by different numerical algorithms. For linear differential equations, one easy way is to use the state-space method mentioned in the previous section.

9.4. DISCRETE-TIME MODELS

Discrete-time modeling and simulation are important in analog system-level design. This is similar to logic simulation for mixed-signal designs, as described in Chapter 5. At the behavioral level, the user describes circuit blocks in terms of **z**-domain transfer functions, difference equations, or by models written in an AHDL. At the functional level, idealized elements are used to represent the circuit elements, and at the circuit level the actual elements are used. Each of these options is explored below.

9.4.1. Behavioral AHDL Models

Discrete-time behavioral models can also be created from an AHDL description using the same procedure used for continuous-time models. The most prominent difference between the two types of models arises from the existence of a master clock for the discrete-time models. The master clock specifies the time-points at which the models will be evaluated. It is appropriate to use discrete-time behavioral models when the behavior of the block is of interest only at certain well-specified points. For example, a discrete-time filter may act on its input data only at certain clocking instants. Therefore, it would be inappropriate to use time-step selection to schedule and evaluate the block at intermediate time-points. Alternatively, the discrete-time model may be evaluated when its own local clock input is high. This corresponds to an enable input for the model which may be driven by an internal node. In this case, the model is very similar to a standard logic block.

9.4.2. Difference Equations and z-Domain Models

Discrete-time models can be represented using difference equations or **z**-domain transfer functions. Difference equations take the following form:

$$\mathbf{y[k+q] + b_1y[k+q-1] + b_2y[k+q-2] + \cdots + b_0y[k] =}$$

$$\mathbf{a_0u[k+q] + a_1u[k+q-1] + \cdots + a_qu[k]}$$

where $\mathbf{u}$ is the input, $\mathbf{y}$ is the output, and $\mathbf{y[k]}$ is defined as $\mathbf{y(kT)}$ with $\mathbf{T}$ as the sampling period. This type of equation can be entered directly into the simulator since it is a time-domain representation of a discrete-time function. A $\mathbf{z}$-domain function takes the form:

$$\mathbf{H(z)} = \frac{\mathbf{Y(z)}}{\mathbf{U(z)}} = \frac{\mathbf{a_0 + a_1 z^{-1} + \cdots + a_p z^{-p}}}{\mathbf{1 + b_1 z^{-1} + \cdots + b_q z^{-q}}} + \mathbf{r} = \frac{\mathbf{N(z)}}{\mathbf{D(z)}} + \mathbf{r} \quad (9.12)$$

where $\mathbf{p < q}$. This equation is the discrete-time counterpart of the formulation described above for $\mathbf{s}$-domain models. Therefore, to map this function to the time-domain, a similar procedure is used. Let $\mathbf{m(k)=m(kT)}$ represent the time response sampled with a sampling period $\mathbf{T}$ and define $\mathbf{q}$ state variables as follows:

$$\mathbf{x_1(k-1) = m(k-q)}$$

$$\mathbf{x_2(k-1) = m(k-q+1)}$$

$$\vdots$$

$$\vdots$$

$$\mathbf{x_{q-1}(k-1) = m(k-2)}$$

$$\mathbf{x_q(k-1) = m(k-1)}$$

and

$$\mathbf{x_q(k) = u(k) - b_q x_1(k-1) - b_{q-1} x_2(k-1) - \cdots - b_1 x_q(k-1)} \quad (9.13)$$

Then, the state equations can be expressed in controllable canonical form as:

$$\begin{bmatrix} \mathbf{x_1(k)} \\ \mathbf{x_2(k)} \\ \vdots \\ \mathbf{x_{q-1}(k)} \\ \mathbf{x_q(k)} \end{bmatrix} = \begin{bmatrix} 0 & 1 & 0 & .. & 0 \\ 0 & 0 & 1 & .. & 0 \\ \vdots & \vdots & \vdots & \vdots & \vdots \\ 0 & 0 & 0 & .. & 1 \\ -\mathbf{b_q} & -\mathbf{b_{q-1}} & -\mathbf{b_{q-2}} & .. & \mathbf{b_1} \end{bmatrix} \begin{bmatrix} \mathbf{x_1(k-1)} \\ \mathbf{x_2(k-1)} \\ \vdots \\ \mathbf{x_{q-1}(k-1)} \\ \mathbf{x_q(k-1)} \end{bmatrix} + \begin{bmatrix} 0 \\ 0 \\ \vdots \\ 0 \\ 1 \end{bmatrix} \mathbf{u(k)}$$

and

$$\mathbf{y(k)} = [\mathbf{a_p} \quad \mathbf{a_{p-1}} \quad \cdots \quad \mathbf{a_1} \quad \mathbf{a_0}] \begin{bmatrix} \mathbf{x_{q-p}(k)} \\ \mathbf{x_{q-p+1}(k)} \\ \vdots \\ \mathbf{x_{q-1}(k)} \\ \mathbf{x_q(k)} \end{bmatrix} + \mathbf{ru(k)}$$

Numerical integration is not required in the case of **z**-domain transfer functions. At each sampling time-point, the values of the state variables and the output value are updated using the canonical form. Note that, as in the case of **s**-domain transfer functions, DC conditions for **z**-domain block must be user-specified to maintain consistency with the other subcircuits. It is not possible to derive DC conditions from the transfer function alone.

9.4.3. Functional Simulation

At the functional level, a charge-conserving, discrete-time algorithm can be used to simulate circuits such as switched-capacitor filters. For example, the methods used in SWITCAP [FAN83] are appropriate here. At this level, a combination of ideal switches, VCVSs and capacitors is sufficient to model linear switched-capacitor circuits. The discrete-time algorithm solves a set of charge-based equations once per clock phase. The underlying assumption is that complete charge transfer is achieved instantaneously after a clock transition. This approach has thc virtue of being several orders of magnitude faster than a SPICE-like approach. The drawback is that second-order effects which introduce signal distortion cannot be easily modeled. In addition, resistive effects are not included in this approach.

We now describe the approach using the simple switched-capacitor integrator circuit shown in Fig. 9.4. The capacitor, $\mathbf{C_S}$, is called the sampling capacitor, and $\mathbf{C_I}$ is the integrating capacitor. A simple two-phase, non-overlapped clocking scheme is used to generate the two ideal switch controls, **Φ1** and **Φ2**. The opamp is modeled as an ideal voltage-controlled voltage source with gain **a**. Charge transfer from the input to

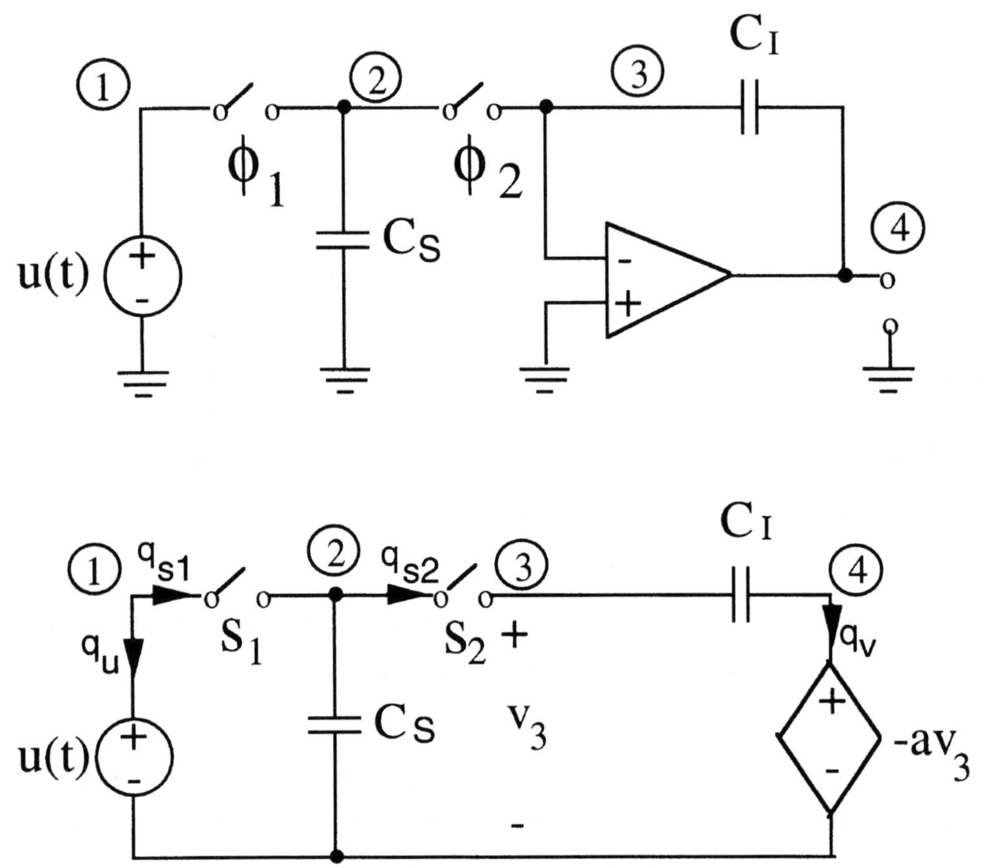

Figure 9.4: SC Integrator

C_S takes place during **Φ1** and charge redistribution occurs between **C_S** and **C_I** during **Φ2**.

An MNA formulation is used to be consistent with the method used in the continuous-time domain. In each of the two phases, charge-conservation equations are written for each node, and branch constitutive relations are added to form the MNA matrix. Let $\mathbf{q_{s1}}$ denote the charge that flows through switch S1, $\mathbf{q_{s2}}$ the charge through S2, $\mathbf{q_u}$ the branch charge through the independent voltage source, and $\mathbf{q_v}$ the branch charge through the VCVS. In phase *k* with S1 closed (**Φ**1 high, and S2 open), the complete set of equations describing the behavior of the circuit can be expressed in terms of an MNA matrix and a vector of unknowns as:

$$\begin{bmatrix} 0 & 0 & 0 & 0 & 1 & 0 & 1 & 0 \\ 0 & \mathbf{C_S} & 0 & 0 & 0 & 0 & -1 & 0 \\ 0 & 0 & \mathbf{C_I} & -\mathbf{C_I} & 0 & 0 & 0 & 0 \\ 0 & 0 & -\mathbf{C_I} & \mathbf{C_I} & 0 & 1 & 0 & 0 \\ 1 & 0 & 0 & 0 & 0 & 0 & 0 & 0 \\ 0 & 0 & \mathbf{a} & 1 & 0 & 0 & 0 & 0 \\ 1 & -1 & 0 & 0 & 0 & 0 & 0 & 0 \\ 0 & 0 & 0 & 0 & 0 & 0 & 0 & 1 \end{bmatrix} \begin{bmatrix} \mathbf{v}_1^k \\ \mathbf{v}_2^k \\ \mathbf{v}_3^k \\ \mathbf{v}_4^k \\ \mathbf{q_u} \\ \mathbf{q_v} \\ \mathbf{q_{s1}} \\ \mathbf{q_{s2}} \end{bmatrix} = \begin{bmatrix} 0 \\ \mathbf{C_S}\mathbf{v}_2^{k-1} \\ \mathbf{C_I}(\mathbf{v}_3^{k-1} - \mathbf{v}_4^{k-1}) \\ \mathbf{C_I}(\mathbf{v}_4^{k-1} - \mathbf{v}_3^{k-1}) \\ \mathbf{u}^k \\ 0 \\ 0 \\ 0 \end{bmatrix}$$

Correspondingly, in phase $k + 1$, with S1 open and S2 closed, the equations are:

$$\begin{bmatrix} 0 & 0 & 0 & 0 & 1 & 0 & 0 & 0 \\ 0 & \mathbf{C_S} & 0 & 0 & 0 & 0 & 0 & 1 \\ 0 & 0 & \mathbf{C_I} & -\mathbf{C_I} & 0 & 0 & 0 & -1 \\ 0 & 0 & -\mathbf{C_I} & \mathbf{C_I} & 0 & 1 & 0 & 0 \\ 1 & 0 & 0 & 0 & 0 & 0 & 0 & 0 \\ 0 & 0 & \mathbf{a} & 1 & 0 & 0 & 0 & 0 \\ 0 & 0 & 0 & 0 & 0 & 0 & 1 & 0 \\ 0 & 1 & -1 & 0 & 0 & 0 & 0 & 0 \end{bmatrix} \begin{bmatrix} \mathbf{v}_1^{k+1} \\ \mathbf{v}_2^{k+1} \\ \mathbf{v}_3^{k+1} \\ \mathbf{v}_4^{k+1} \\ \mathbf{q_u} \\ \mathbf{q_v} \\ \mathbf{q_{s1}} \\ \mathbf{q_{s2}} \end{bmatrix} = \begin{bmatrix} 0 \\ \mathbf{C_S}\mathbf{v}_2^k \\ \mathbf{C_I}(\mathbf{v}_3^k - \mathbf{v}_4^k) \\ \mathbf{C_I}(\mathbf{v}_4^k - \mathbf{v}_3^k) \\ \mathbf{u}^{k+1} \\ 0 \\ 0 \\ 0 \end{bmatrix}$$

Of course, in practice there may be more than two phases, depending on the specific clocking scheme used. Since the MNA matrices remain unchanged in each phase, each such MNA matrix can be computed once and then cached for later reuse. The matrices can be stored in an LU-decomposed form for greater efficiency. To avoid excessive use of memory, a limit should be set on how many such matrices can be cached, especially when there are a large number of phases.

The simulation interval is usually made up of hundreds, or possibly thousands, of clock cycles. During a transient simulation, one MNA matrix is evaluated in every clock cycle and a new phase is signaled by a clock transition. The program checks the switch settings in the new phase against stored values to determine whether this setting has been encountered before. If it has, the appropriate MNA matrix is accessed and used to solve the system of linear equations. Otherwise, a new MNA matrix is created from element stamps. This process continues until all the phases have been processed.

9.5. MIXED CONTINUOUS/DISCRETE SIMULATION

It is clear that a complete analog multilevel simulator should provide all the levels shown in Fig. 1.1 and perform both discrete-time and continuous-time simulation. In fact, for some circuits such as the oversampled A/D converter shown in Fig. 9.1, a mixed continuous/discrete simulation capability is essential for efficient simulation. The filter at the front end could be simulated using an **s**-domain transfer function or at the electrical level. The modulator, which is usually implemented as a switched-capacitor circuit, could be modeled at the functional level in the discrete-time domain. Finally, the decimation filter could be described in terms of a **z**-domain transfer function or as a discrete-time behavioral block. This combination of discrete-time and continuous-time models is well-suited to the natural structure of the circuit and would provide functional verification and first-order timing information in a short time.

Table 9.2 shows the combined simulation hierarchy that would be needed in a true mixed continuous/discrete simulator. There are two basic issues that must be considered when designing this type of simulator. The

	Continuous Time	*Discrete Time*
Behavioral Level	s-domain functions, differential equations and blocks described in AHDL	z-domain functions, discrete-time equations and blocks described in AHDL
Functional Level	Nonlinear and linear controlled sources, Logic gates, ADC, DAC, opamps etc.	Voltage-controlled switches, voltage source
Circuit Level	Transistors, Diodes, Capacitors, Resistors, Inductors	Capacitors, Transistors

Table 9.2: Mixed Continuous/Discrete simulation

first is the design of a mechanism to maintain a consistent solution for the circuit at any given time point in the simulation interval. In the case of a continuous-time subcircuit, the inputs and outputs are of interest at every timepoint in the interval considered. It is inherent in the time-point selection that the output value at any intermediate time can be interpolated from the values at the time-points. However, for a discrete-time subcircuit the inputs and outputs are of interest only at certain discrete-time instants associated with some clocks or sampling processes. The simulator must convey the appropriate information from continuous-time to discrete-time subcircuits (and vice-versa) as needed during the simulation process, and synchronization must be maintained between the different subcircuits.

The second issue concerns convergence to a solution in the presence of discontinuities. An element such as an ideal switch can have a very sharp transition at its output when it turns on. In fact, the voltages at its output terminal can change almost instantaneously. In some cases these voltages may be the inputs to continuous-time electrical-level or functional-level subcircuits. The integration methods used at the electrical and functional levels are extremely prone to convergence problems and numerical errors in the presence of sharp transitions or discontinuities. Furthermore, the local truncation error checking schemes used in conjunction with the integration methods have to be modified. This is due to the fact that voltages at the affected circuit nodes prior to the transition cannot reliably be used as predictors of the voltages at the circuit nodes after the transition, since the relative voltage change per unit time may be large.

An event-driven paradigm can be used to maintain consistency of the solution in the two domains. It would operate as follows. A given circuit is partitioned by the simulator into discrete-time and continuous-time subcircuits, based on the models specified by the user. For a continuous-time subcircuit, an event is defined to occur when the state of one or more of its nodes changes, as described Chapter 4. When an event occurs, a subcircuit will try to schedule all of its fanout subcircuits for processing by the simulator. As shown in Fig. 9.5, a continuous-time subcircuit can fanout to either discrete or continuous-time subcircuits. With reference to the figure, a continuous-time subcircuit can schedule another continuous-time subcircuit but a discrete-time subcircuit can only be scheduled by a specially designated clock input. On the other hand, whenever the output of a discrete-time subcircuit changes, it can schedule any continuous-time subcircuit on its fanout list. This mechanism ensures that a discrete-time subcircuit will not be scheduled at an inappropriate time and also permits

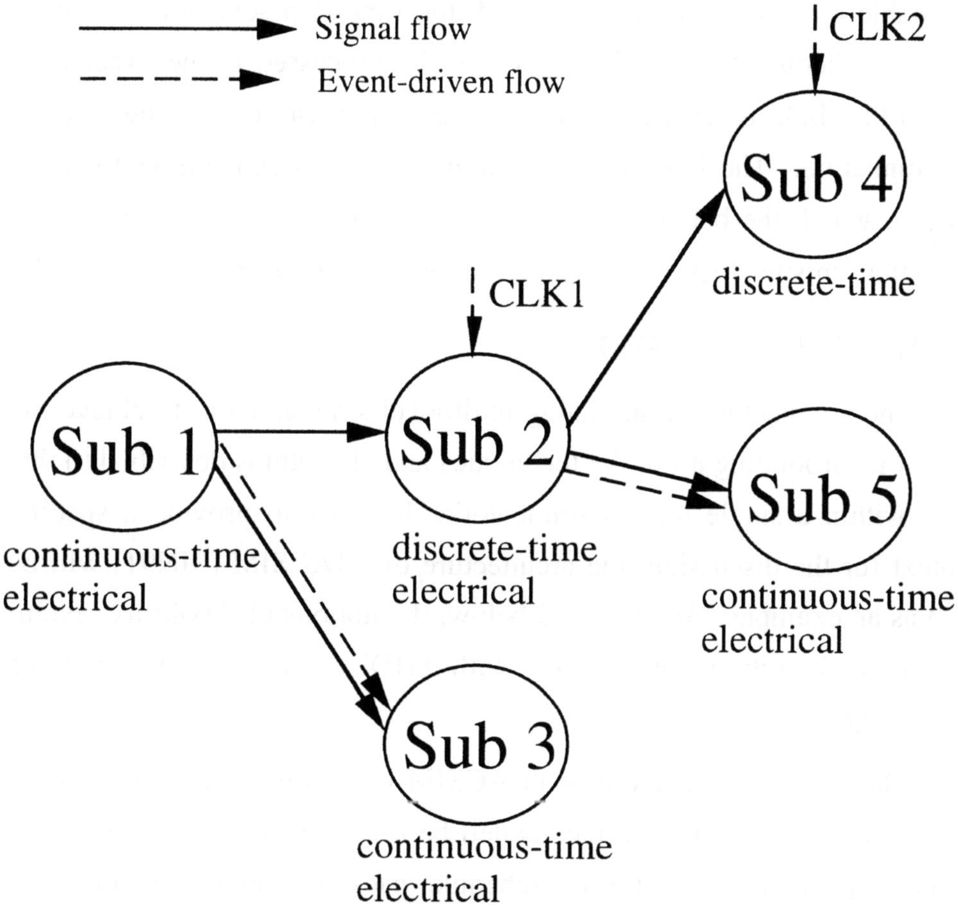

Figure 9.5: Combined Simulation of Continuous and Discrete Time Structure

latency in the circuit to be exploited.

Another important issue that must be considered is that a clocking signal for a discrete-time subcircuit may not be known in advance. The user should have the option of either completely specifying the clocking signal in advance before the simulation, or deriving it from internal nodes in the circuit, whose behavior is not known a priori. In general, this problem may exist even in purely continuous-time circuits containing elements whose behavior changes at some threshold (usually a threshold voltage). If suitable constraints are not imposed, the time-step selection algorithm may not schedule a threshold element to be processed at the exact time that its threshold is reached, and this would give rise to a timing error in the simulation. The issue of determining the exact scheduling time for cases in which the behavior of the clocking signal is not known a priori (nondeterministic clocks) is discussed in more detail in [BED91, SIN94B].

9.6. iMACSIM: A CASE STUDY

The architecture of an analog multilevel simulator must achieve the goal of incorporating all of the techniques and algorithms described to this point within a single unified framework. In order to provide a specific context for the discussion, the architecture of iMACSIM [SIN91] will be used as an example. As described below, the notions of flexibility, extensibility, modularity and ease-of-use with AHDLs were key considerations in its design.

The overall architecture of iMACSIM is shown in Fig. 9.6. It consists of a programming interface called he *algorithmic backplane*, which contains sparse matrix routines, a scheduler package, waveform processing routines, and input/output routines. Several different algorithms are shown that plug directly into the backplane. Each algorithm has a corresponding set of models dedicated to it. A given model will be processed only by its associated algorithm. This modular architecture enables new algorithms or models to be easily added or removed, providing the program with a high degree of adaptability. The algorithmic backplane is a procedural interface for the simulation algorithms. It consists of a set of macros and subroutine calls that allow the algorithm developer to access

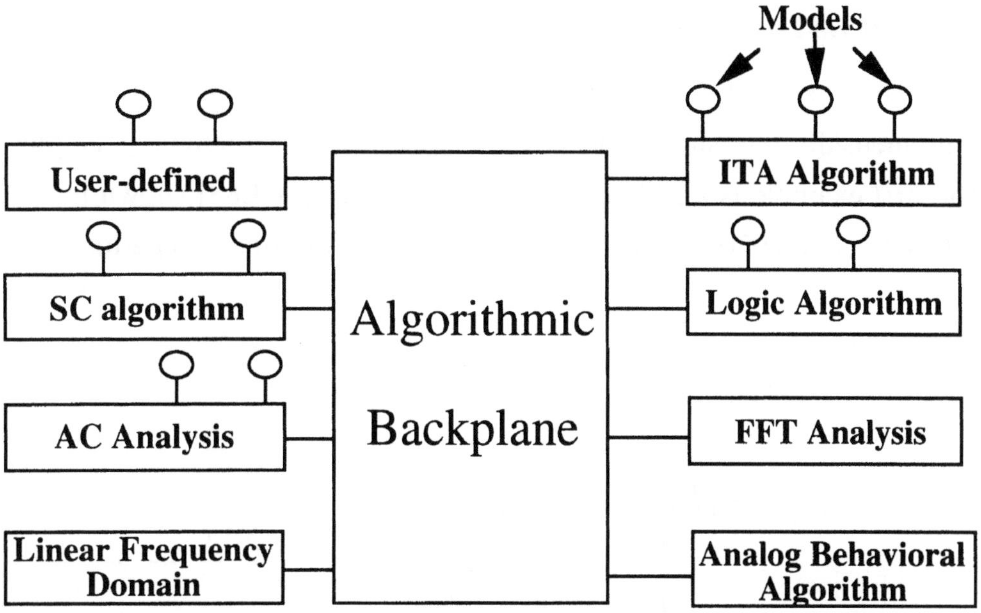

Figure 9.6: Simulator Architecture

data structures and operations without a detailed understanding of the rest of the program.

Fig. 9.6 shows some of the simulation algorithms which are already attached to the backplane. The iterated timing analysis (ITA) algorithm, described in Chapter 4, performs electrical level simulation. The analog behavioral algorithm controls both discrete-time and continuous-time models. The associated models include **s**-domain functions and continuous-time blocks, and **z**-domain functions and discrete-time blocks. The logic algorithm processes behavioral descriptions of logic gates, as described in Chapter 5. The SC algorithm processes switched-capacitor subcircuits in the time-domain, as described earlier in this chapter. Since iMACSIM uses an event-driven paradigm, the inner loop of the program is an event processor which is identical to that described in Chapter 8.

9.7. SIMULATION EXAMPLES

This section illustrates multilevel simulation and mixed continuous/discrete simulation in iMACSIM using two simple examples. The first example is a PLL-based clock generator [SHE88] shown in Fig. 9.7 that contains 205 MOS transistors. To speed up the simulation time,

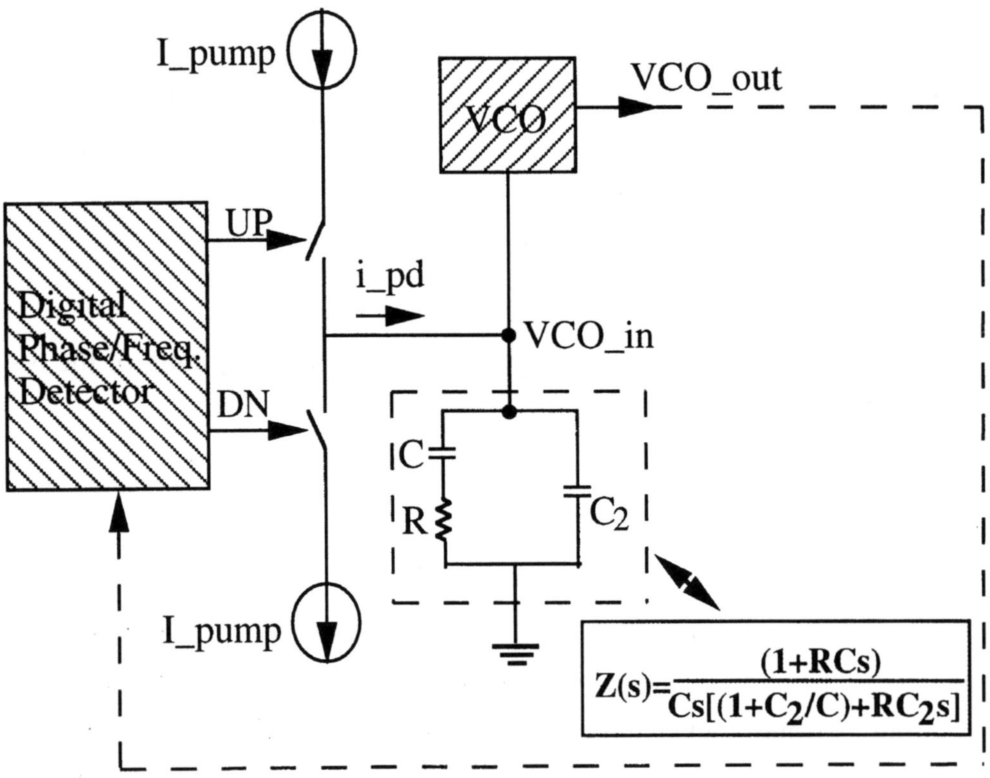

Figure 9.7: Charge Pump, Filter and VCO of Clock-Generator

the circuit was represented using an **s**-domain model and behavioral logic models. The low-pass filter in the PLL has the impedance transfer function:

$$\mathbf{Z}(\mathbf{s}) = \frac{1 + \mathbf{RCs}}{\mathbf{Cs}((1 + \mathbf{C}_2/\mathbf{C}) + \mathbf{RC}_2\mathbf{s})} \quad (9.14)$$

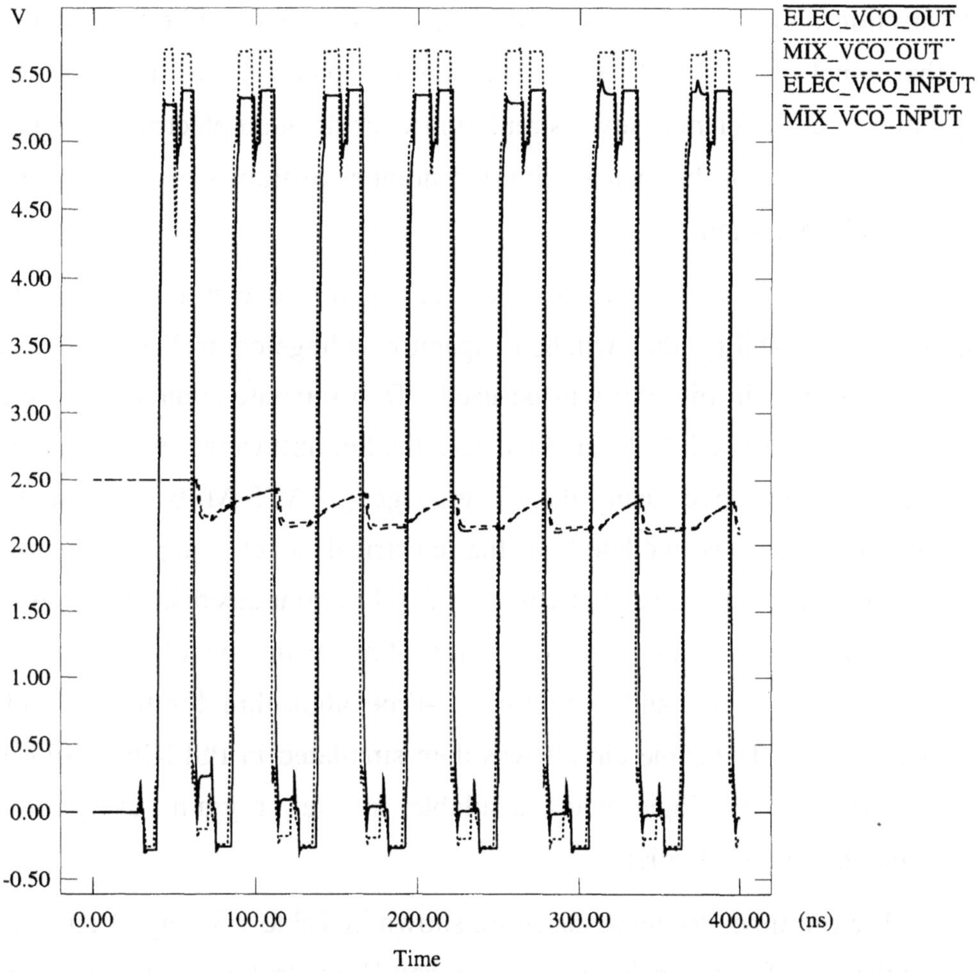

Figure 9.8: Comparison of Voltage at the VCO Input and Output

and was represented by an s-domain block which has a single bidirectional connection to the rest of the circuit.

The digital phase detector was modeled using 40 logic gates in its behavioral description. To maintain a high level of accuracy, it was important to keep the nonlinear VCO at the electrical level, and 95 transistors from the original electrical description were used to model it. Fig. 9.8 compares waveforms obtained from the electrical and multilevel circuit descriptions. One waveform is at the input of the VCO while the other is at the output of the VCO. The results are comparable, and the observed speedup is 2.5X for multilevel simulation over detailed electrical simulation. This is reasonable considering the number of transistors used in the multilevel simulation.

To demonstrate the effectiveness of the mixed continuous/discrete simulation capability, the switched-capacitor voltage-controlled oscillator [HOS84] shown in Fig. 9.9 will be used. Two separate simulations were conducted in iMACSIM. In the first case, the Schmitt trigger was modeled at the behavioral level using the **C** language (iMACSIM_B), and in the second case it was modeled at the electrical level using transistors (iMACSIM_E). In the second case, a mixed continuous/discrete simulation was performed: a discrete-time algorithm for the switched-capacitor portion of the circuit and a continuous-time algorithm for the Schmitt trigger circuit. The same circuit was then simulated in PSPICE [PSP90], with switches and logic gates available in the program, using only continuous-time simulation.

The run times for these cases are shown in Table 9.3. Fig. 9.10 compares the waveforms for the two cases when V_control is swept from -5 to +5 V. In accordance with the results presented in [SUY89], the oscillation frequency changes from 500 Hz to 1500 Hz. iMACSIM, with a transistor level description for the Schmitt trigger (row 2) was 18X faster than

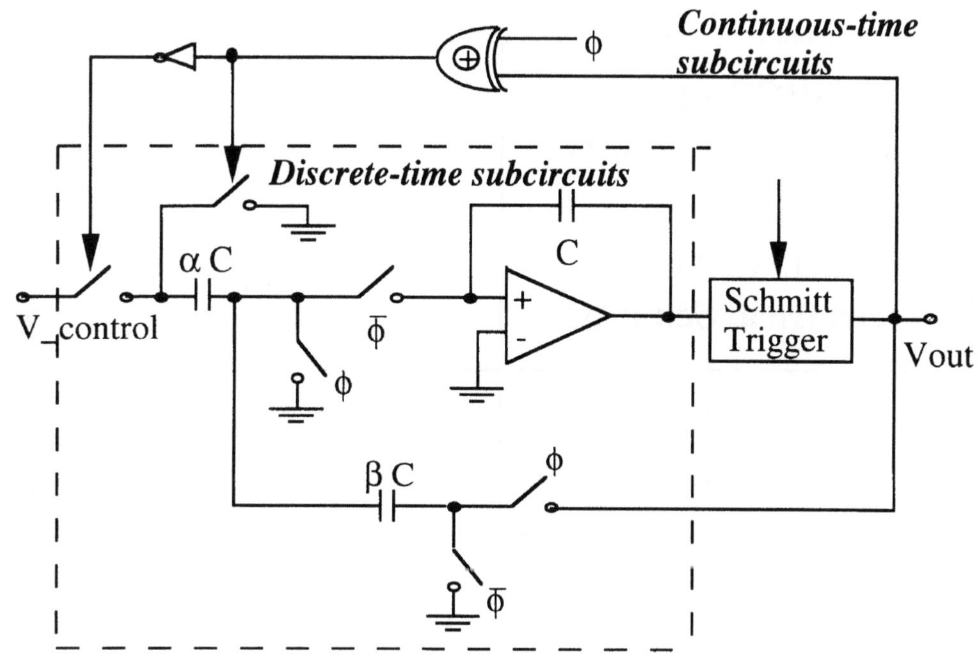

Figure 9.9: Switch-Capacitor Voltage-Controlled Oscillator

PSPICE (row 1). The result for the behavioral Schmitt trigger was 10X faster than the electrical Schmitt trigger. The results from the two simulations are not identical since the accuracy of the behavioral model of the Schmitt trigger is limited. However, this example demonstrates that good speedups and acceptable accuracy can be obtained using a mixed continuous/discrete approach.

Program	Run-time(sec)
PSPICE v4.05 (electrical Schmitt)	2540
iMACSIM (electrical Schmitt)	136
iMACSIM (behavioral Schmitt)	13

Table 9.3: Run Time for VCO on Sun SPARC2

9.8. A MACROMODELING AND SIMULATION ENVIRONMENT

We now describe various techniques for macromodeling and present a complete system for analog macromodeling and simulation. A brief discussion of the macromodeling process was presented earlier in this chapter for linear models, but general macromodeling issues were not described. The essence of macromodeling is to capture the important input/output characteristics of a complex circuit in a simplified model called a macromodel. The simulation is then performed using a number of these macromodels to reduce the overall simulation time. This has also been referred to as model order reduction.

There are a number of ways to generate macromodels for a given

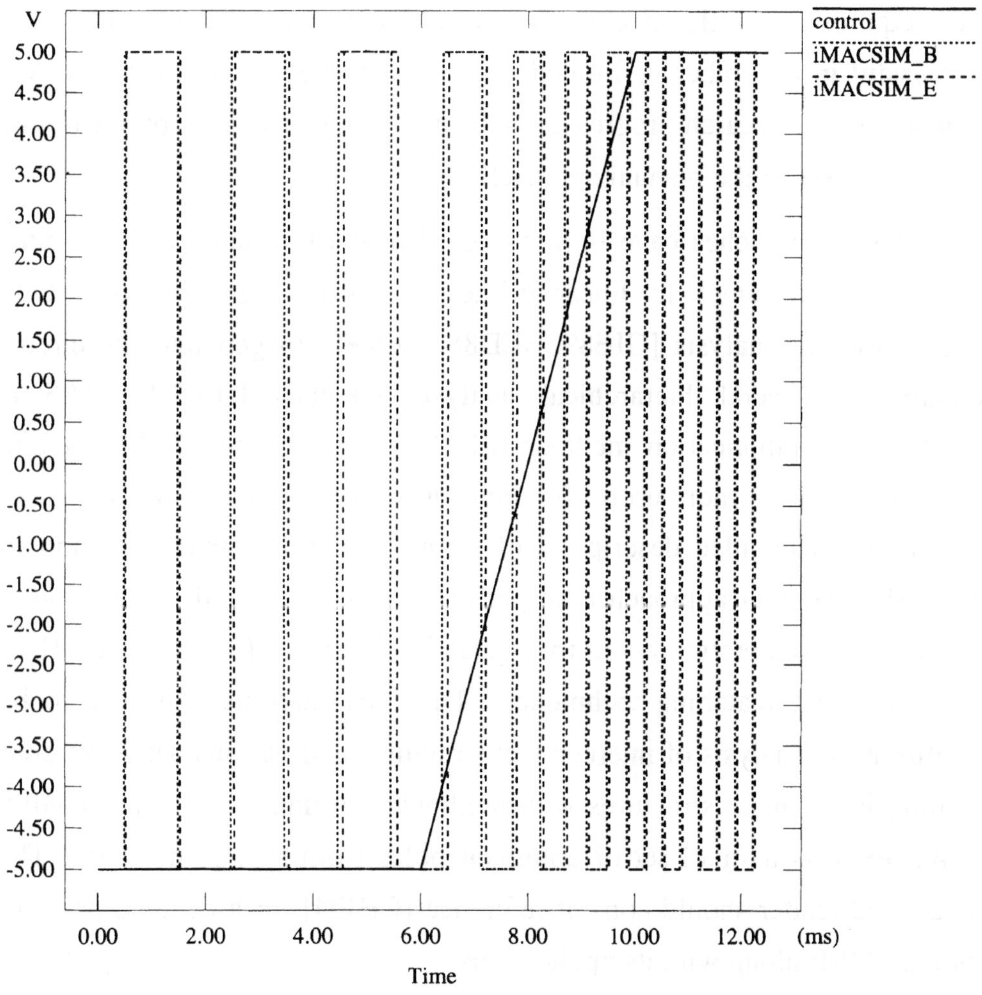

Figure 9.10: Oscillator Outputs as A Function of V_control

transistor level circuit block. The manual approach has been used for many years and relies on the knowledge and intuition of the designer. This has proven to be very successful, although there is an enormous amount of work involved in developing a complex macromodel that includes all of the desired effects and is accurate across different applications. However, once it has been developed, the runtime advantages of using a macromodel are also enormous. Automatic macromodel generation requires very little effort on the part of the designer, but has only been successful in a number of limited situations, namely, linear circuits. Of course, as the research progresses, it is anticipated that this approach will be preferred over the manual approach.

Two approaches have been explored recently to automatically generate the macromodels for linear or linearized (small-signal) circuits. One uses symbolic analysis [GIE89, SED88, FER91] to generate the circuit equations. Several hierarchical analysis methods [HAS89, HAS91, JOU94] and simplication methods [FER92, SED92, HSU93] have made symbolic analysis more useful for applications such as circuit analysis and synthesis. A detailed description of symbolic analysis may be found in [GIE91]. Another approach is to perform model order reduction through the use of asymptotic waveform evaluation (AWE) [PIL90, RAG93] or other moment matching techniques. These methods have proven to be useful in the analysis of interconnect structures and various networks containing large linear structures with nonlinear terminations. Conceptually, AWE extracts a small set of dominant poles from a large network. The interested reader should consult reference [CHI94] for a detailed description of AWE along with its applications.

The macromodeling process in the nonlinear case is mainly manual at present. Little progress has been made in the automatic generation of nonlinear macromodels. One disadvantage common to most of the high-level simulators available today is the large effort required on the part of the designer to create a new macromodel. To circumvent this problem, a large library of useful macromodels usually accompanies most commercial simulators. However, if a needed macromodel is not available in the library, the long and tedious process of creating a new one must be undertaken.

For circuit designers who do not have a CAD background, creating the necessary macromodels can be a frustrating experience. Usually the designer must have programming skills in a language such as C or FORTRAN in addition to possessing a good working knowledge of the simulator. To compound the problem, newly developed primitives often have numerical problems that the designer is unaware of such as nonconvergence, discontinuities, numerical overflow, and so on. It would be useful to identify these problems before the macromodels are used in an actual simulation. In order to automate this entire process, the designer should be able to specify a macromodel in a hardware description language that is specifically tailored for analog designs. The designer's description should be checked for potential numerical problems without the designer needing to be aware of the internal workings of the simulator [MA92, CHA92, VIS88]. Finally, the macromodel should be automatically optimized to deliver the intended performance.

A suite of programs has been developed at the University of Illinois to serve as a vehicle for demonstrating these concepts. It includes an AHDL language translator called iMacGen [MA92], a numerical consistency checker called iMacChk [MA92] and an optimizer called iMAVERICK [JU91]. iMacGen and iMacChk have been tailored to interface with iMACSIM. Ideally, a fully automated macromodeling tool would automatically create the desired macromodel, incorporate the necessary primitives into the simulator, and then optimize any macromodel parameters to match the behavior of the original circuit. However, a system with these features is rather unrealistic at this point in time. Instead, these tools address the most tedious and time-consuming portions of macromodel, and leave the creative part in the hands of the designer.

Fig. 9.11 shows the overall structure of the macromodeling process using these tools. First, the designer identifies the circuit block to be replaced by a macromodel and constructs a macromodel using primitives. The primitives that do not already exist can be added to the simulator in an AHDL format through iMacGen. The integrity and functionality of the newly created primitives must be checked before they can be safely used. For this purpose, iMacChk is used to detect some of the common problems that the primitives might have in the user-specified region of operation. After all the primitives have been entered and checked, a macromodel with a set of adjustable parameters can be constructed. The optimizer, iMAVERICK, is used to fine tune the macromodel parameters to the designer's specifications.

Fig. 9.12 provides an overview of iMacGen. This program accepts a model description provided by the designer in an AHDL format, checks its syntax, and then generates a device model in **C** for the iMACSIM program. The syntax of the language is based on combination of VHDL and **C**. The VHDL constructs provide structure for the definition of the new model while the **C** language is used for equation specification. The equations are extracted and passed to Mathematica [WOL91], a well-known symbolic analysis package, so that the derivatives needed by the simulator can be generated. These derivatives, along with the element equations, are used to build the matrix stamp of the device. The input/output information is used to build scheduling tables for the device. The parameters, their default values, and their range of permissible values are converted into initialization and bounds checking routines. Finally, a few tables are updated

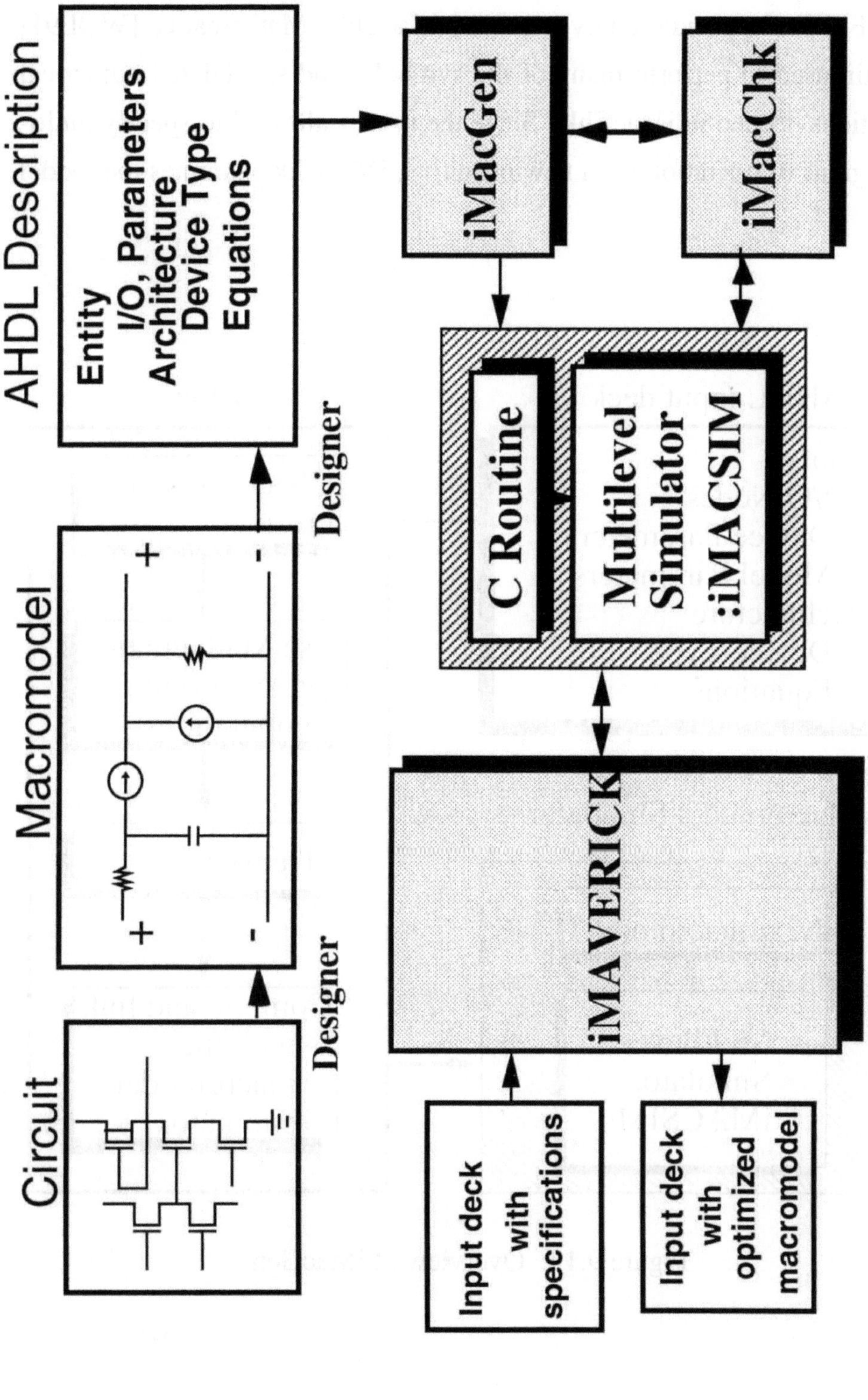

Figure 9.11: Overview of the Entire System

to make the new primitive a recognized element for iMACSIM users.

Fig. 9.13 provides an overview of iMacChk. Mathematica [WOL91] is again used to perform many of the symbolic and specialized numerical operations needed in iMacChk. Since the user is allowed to specify multiple regions of operation for a new primitive, iMacChk examines the model

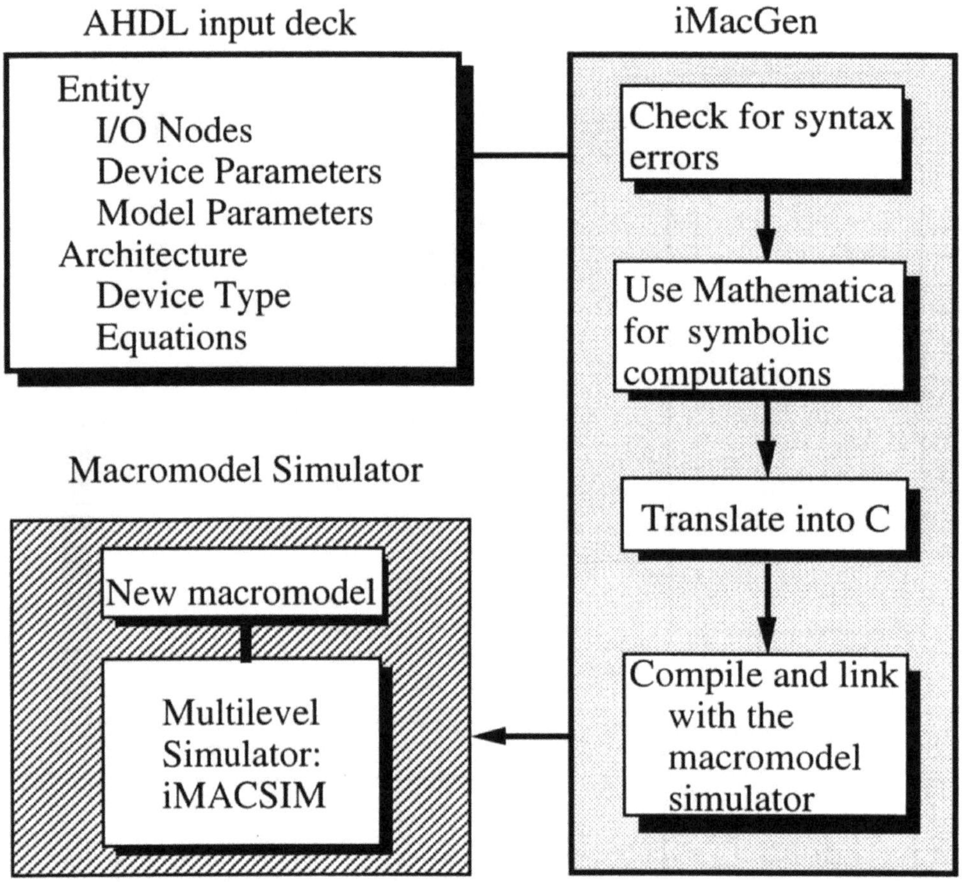

Figure 9.12: Overview of iMacGen

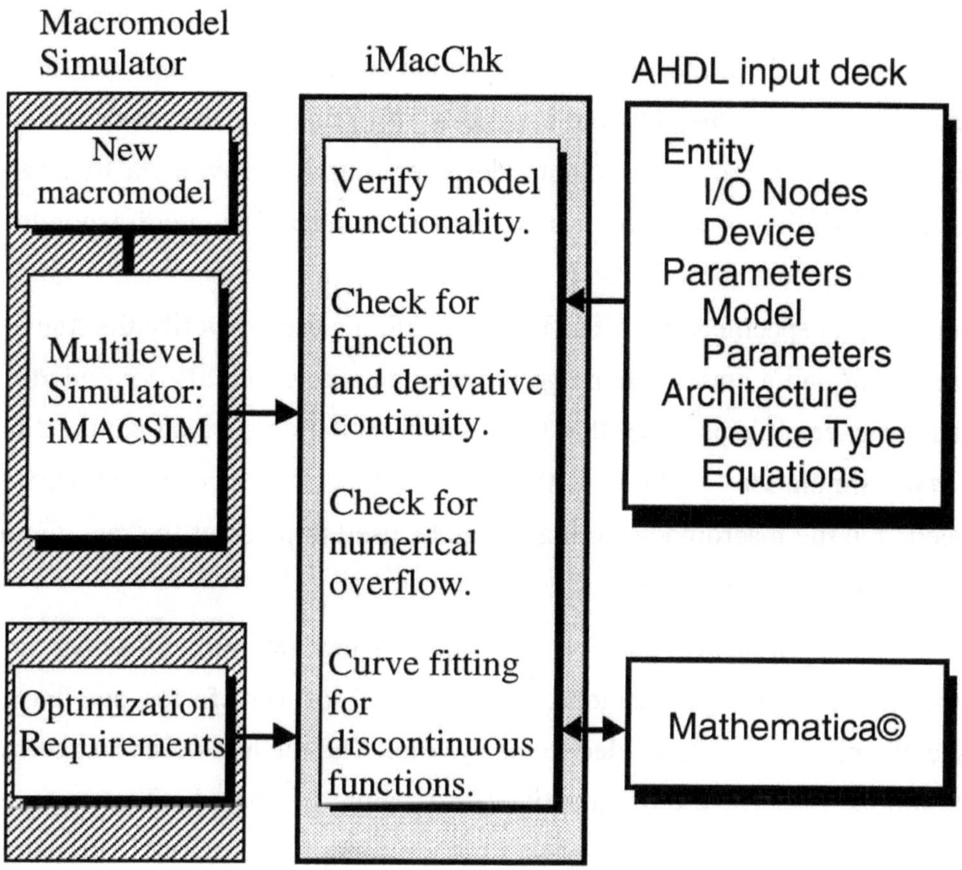

Figure 9.13: Overview of iMacChk

function and its derivatives for continuity by numerically integrating the function and its derivatives across the user specified regions of operation. At all breakpoints, the left and right limits of the function and its derivatives are checked for equality. It also detects potential overflow/underflow problems in the model by systematically sampling the function and identifying the regions in which the value is undefined. iMacChk contains signal-handlers to identify and report numerical overflow. At the present time, the user is required to correct any problems identified by the program.

The iMAVERICK system, as shown in Fig. 9.14, is a closed-loop verification and optimization system, which not only checks the validity of macromodels, but also has the capability of optimizing the macromodels to improve accuracy. The circuit designer initially provides a transistor level description of the circuit block under consideration, a corresponding time-domain macromodel with a number of adjustable parameters, and a set of target specifications. The input excitations used to verify the transistor circuit and a reasonable range for each parameter are also provided by the designer. It is assumed that these input excitations, which are used for circuit simulation, capture all the important performance characteristics needed in the macromodel so that it can be used in place of the real circuit in a particular application.

The optimization can be performed using any mix of scalar and waveform target specifications. Any number of waveforms and target specifications can be provided to the system. The method described below assuming that waveform quantities are being compared. Two sets of waveforms, generated by a circuit simulator and a macromodel simulator, respectively, are used as input to a waveform consistency checker. The output waveforms can be time-domain and/or frequency-domain responses of the circuit. The consistency checker performs a set of comparisons between two sets of waveforms and returns a value indicating the relative proximity of the waveforms. If the macromodel does not compare favorably with the transistor level circuit based on these waveforms, the parameters are adjusted and the simulation and consistency checking cycle is repeated. This optimization process continues until the desired level of accuracy is achieved or the maximum allowable CPU-time is exceeded.

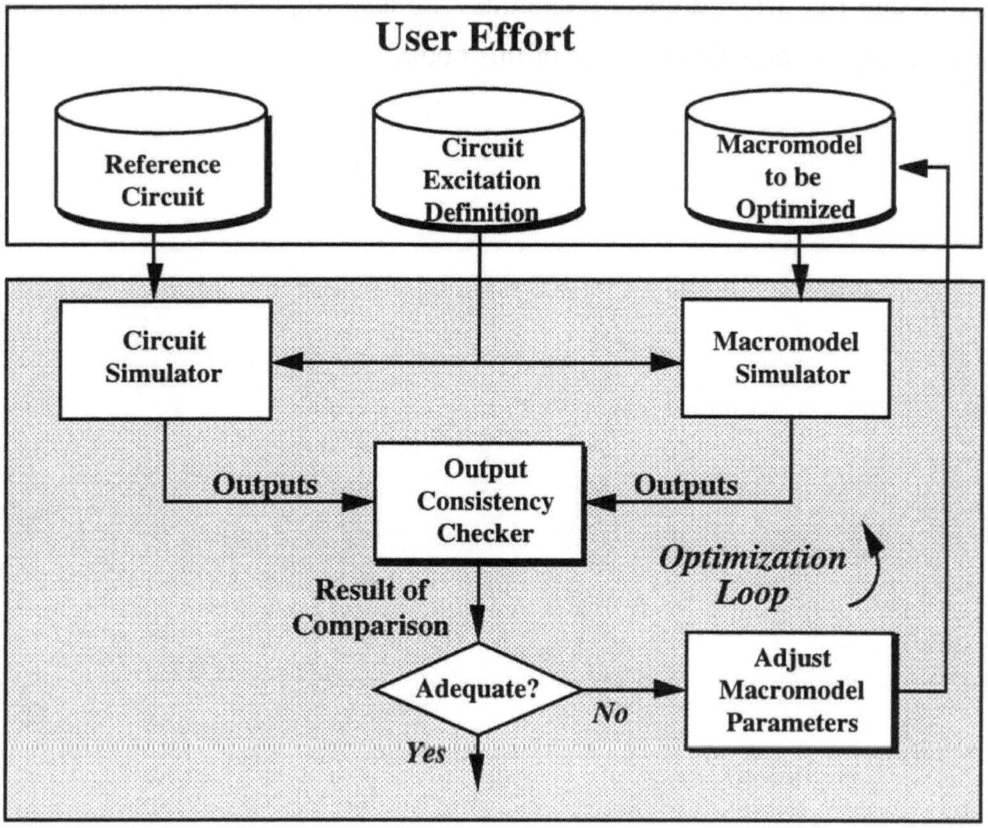

Figure 9.14: Overview of iMAVERICK System

9.9. SUMMARY

This chapter has focused on the extension of analog simulation techniques to include multilevel simulation and mixed continuous/discrete simulation. With the full analog modeling and simulation hierarchy in place, it is now possible to mix and match all the levels shown in Fig. 1.1. For mixed-signal simulation, the circuit can now be described at many levels of abstraction both on the digital side and the analog side. Tools for macromodeling, verification and optimization have been described to assist with this process on the analog side. Together, virtually any form of time-domain simulation of entire mixed-signal systems is now possible. Although frequency-domain techniques are also important for analog circuits, they are outside the scope of this book.

CHAPTER 10

CONCLUSIONS AND FUTURE WORK

10.1. SUMMARY

A variety of techniques for mixed-mode and analog multilevel simulation have been described in this book. The primary focus in mixed-mode simulation was the combination of the gate-level, switch-level timing and electrical forms of simulation. In analog multilevel simulation, the mixing of continuous-time simulation with discrete-time simulation was the primary focus.

Chapter 1 began with an overview of the different levels of simulation and provided the motivation for combining two or more levels into one simulator. Then, the basic issues in mixed-mode simulation were outlined, and a brief survey of existing mixed-mode simulators was provided. In Chapter 2, the electrical simulation problem was formulated and standard numerical techniques used to solve the problem were presented. The issues associated with the implementation of an efficient time-step control scheme were also described. In Chapter 3, two properties of waveforms, called latency and multirate behavior, were defined and used to motivate the need for new circuit simulation methods. The relaxation-based electrical simulation methods were introduced to exploit these waveform properties, and their theoretical aspects were described. Circuit partitioning methods to improve the convergence speed of relaxation methods were presented at the end of the chapter.

The electrical, gate-level and switch-level timing simulation algorithms were presented in Chapters 4, 5 and 6, respectively. These techniques make use of the event-driven, selective-trace paradigm which forms a common thread for all algorithms used in mixed-mode simulation. The main contribution of Chapter 4 was an event-driven circuit simulation algorithm to exploit latency, and a new partitioning algorithm to ensure the convergence of Gauss-Seidel-Newton relaxation in the presence of MNA elements. The evolution of logic states and logic delay models was presented in Chapter 5. The development of the Elogic simulation and modeling approach was described in Chapter 6.

Chapter 7 addressed the important issue of the mixed-mode interface. Signal conversions from the analog domain to the digital domain and vice-versa were described. The problem of **X** state handling that faces most commercial programs was also addressed. Examples were used to illustrate potential problems of the various interface models in use today.

In Chapter 8, the implementation details of a mixed-mode simulator were presented. First, the overall architectural issues were described, followed by a summary of the transient analysis techniques used and event scheduling policies enforced between the different levels of simulation. The issues associated with the implementation of event schedulers were also described in detail. A technique for the dc solution of mixed-level circuits was outlined. Next, the techniques for automatic mixed-mode partitioning were detailed. Finally, a mixed-mode simulation example was provided to show that large speedups could be obtained without sacrificing accuracy.

Chapter 9 was devoted to the simulation problems of analog multilevel simulation. After motivating the need for this type of simulation, the requirements of analog multilevel simulation based on circuit-related issues were developed. Then, a number of commercial simulators were surveyed. Next, the simulation techniques for continuous-time behavioral models and discrete-time models were described in detail. This was followed by the implementation issues associated with the design of the iMACSIM simulator. Finally, the concepts embodied in a complete macromodeling and simulation environment were outlined.

10.2. AREAS OF FUTURE WORK

Although a substantial amount of work has been done in mixed-mode simulation and analog multilevel simulation, there are still many promising areas of future work. In particular, simulation of coupling effects in mixed-signal ICs and the development of an analog hardware simulation language are important areas of research in the near future. These topics are outlined below.

10.2.1. Coupling Effects in Mixed-signal ICs

One problem not addressed in this book is the simulation of the *coupling effects* in mixed-signal ICs. Parasitic coupling of digital switching noise to analog circuits on the same chip through direct capacitive coupling and interaction via the common substrate may corrupt analog signals and degrade the performance of mixed-signal ICs [VER93, MAS92]. Thus, accurate simulation of the effects of digital switching noise coupling into analog nodes, and crosstalk between analog nodes are required to determine the true performance of mixed-signal ICs.

Traditionally, this problem has been minimized by employing conservative layout design, and any problems were addressed after fabrication of the chip. Now, due to the advance in mixed-mode simulation, along with new model generation and reduction techniques, the coupling effects of a circuit block or even the entire circuits may be simulated in a reasonable amount of time. Other sources of problems in mixed-signal designs include VDD bounce, ground bounce, and IR drops along the power lines. These topics have also been the subject of recent research in this area.

10.2.2. Analog Hardware Description Languages

As mentioned earlier, there are currently two major efforts underway towards the standardization of analog hardware description languages (AHDLs) [SAL94A]. Generally speaking, *MHDL* is cast as a solution for circuits with high analog content (including purely analog circuits) while *VHDL-A* provides a uniform environment for predominantly digital circuits with some analog functions. The languages provide constructs to represent the structure and behavior of the design. Hierarchy and inheritance are necessary attributes of these languages that allow a system to be represented at various levels of abstraction. Behavior is expressed through model constructs and structure is represented through connectivity and hierarchical decomposition mechanisms.

The outcome of the standardization efforts will have a significant impact on the way analog design is carried out in the future. It is generally agreed that a standard AHDL is urgently needed in the analog design community. What is not so clear is the form of AHDL that would be widely accepted by tool vendors and analog designers, who will eventually have to represent their designs using the language. Furthermore, the scope of the language in not generally agreed upon. In order to capture all of the necessary features, an AHDL would have to be comprehensive in its coverage as it may be used in synthesis, testing, documentation, and data exchange in addition to simulation. Such a language would be difficult to learn quickly and would require an enormous effort to implement for the vendors. If the scope of the language were limited, its lifetime would be rather limited. It is clear that an intermediate position must be taken to establish an AHDL that could serve the present needs of the analog community, with the potential to grow as the needs change in the future.

Since analog HDLs allow the interconnection of instances of a model in any fashion, this presents a problem when devices represented in different domains are connected at the same node [SAL94A, SIN94]. A worst-case interconnection is shown in Fig. 10.1. Of course, an HDL is not responsible for resolving the problems due to this connection, but merely to give a precise and unambiguous meaning to the description. So this type of interconnection is permissible but the meaning is open to interpretation. In fact, there may be several different "resolutions" for the same connection depending on the user and the tool reading the description. For example, if a time-domain simulation is being requested, all frequency domain models would be converted to the time domain before the analysis begins. This conversion could be performed automatically, or an error message generated if the conversion is not possible. The resolution of all of the conflicts created by the connection shown in Fig.10.1 should provide fertile ground for future work in AHDL.

10.3. CONCLUSIONS

Mixed-mode simulation and analog multilevel simulation are now well-accepted forms of simulation in industry for mixed-signal ICs, and for analog circuits described at multiple levels of abstraction. A wide variety of simulators have been developed, both in industry and academia, and many are in use today. As described in this book, the key contribution of mixed-mode and analog multilevel simulation is that they offer the designer the ability to intelligently trade off simulation and modeling precision for simulator performance within the scope of a single simulator. This permits the designer to choose detailed simulation where accuracy is essential and higher forms of simulation where less accuracy can be tolerated.

A second important theme of the book is that multilevel simulators provide a uniform environment for designers to develop ideas from initial concepts to the final circuit schematics and accommodates both top-down

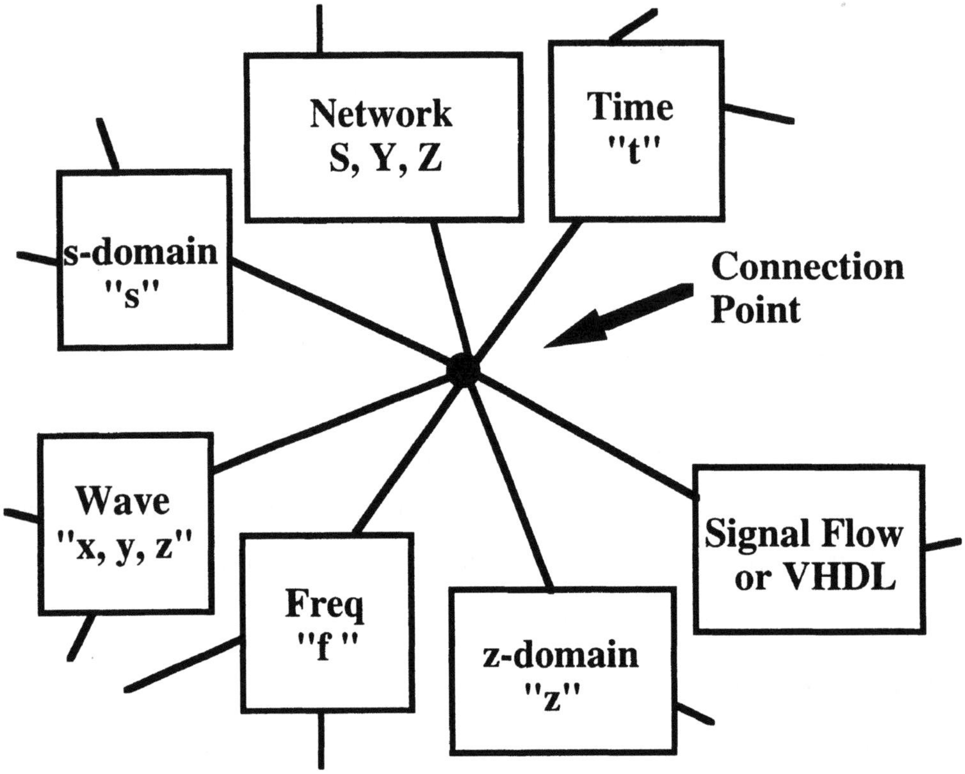

Figure 10.1: "Worst Case" Interconnection

and bottom-up design styles, or any form in between. Designers can focus on system level or architectural issues in the preliminary phase of the design and progressively add more detail as design decisions are made at each level of abstraction. In addition, designers can mix and match different levels of abstraction in a single schematic diagram to convey the important aspects of a circuit design. These different representations can be captured easily in a mixed-mode and analog multilevel simulation environment and later used to verify the circuit operation and performance. Finally, these types of simulators are flexible and extensible and provide high performance in circuit verification. These features combine to place them among the most important tools in VLSI design.

REFERENCES

[ACU89] E. Acuna, J. Dervenis, A. Pagones, R. Saleh, "iSPLICE3: A New Simulator for Mixed Analog/Digital Circuits", Custom Integrated Circuits Conference Digest of paper, May 1989, pp. 13.1.1-13.1.4.

[ACU90] E. L. Acuna, "Simulation, Partitioning, and Characterization of Logic Blocks in Mixed-Mode Simulation", M.S. Thesis, University of Illinois at Urbana-Champaign, 1990.

[AGR80] V. D. Agrawal, A. K. Bose, P. Kozak, H. N. Nham, "A Mixed-Mode Simulator", Proc. of 17th Design Automation Conference, June 1980, pp. 618-625.

[ANA93] ELDO User's Manual, ANACAD, June 1993.

[ANT93] B. A. A. Antao, "Synthesis and Verification of Analog Integrated Circuits", Ph.D Dissertation, Vanderbilt University, Dec. 1993.

[ARN78] G. Arnout, H. DeMan, "The Use of Thresholding Functions and Boolean-Controlled Elements for Macromodelling of LSI Circuits", IEEE J. of Solid-State Circuits, SC-13, June 1978, pp. 326-332.

[AT&91] "The ABCDL: A Robust Environment for Analog Circuit Behavioral Modeling", AT&T Bell Laboratories Internal Technical Memorandum, March 1991.

[BEA88] D. L. Beatty and R. E. Bryant, "Incremental Switch-Level Analysis", IEEE Design and Test of Computers, Vol. 5, December 1988, pp. 33-42.

[BED91] D. G. Bedrosian, "Analysis of Networks with Internally Controlled Switches", Ph.D Thesis, University of Waterloo, 1991.

[BEN91] J. Benkoski, J. Besnard, S. Gai, M. Magni, E. Profumo, "Mozart_MM: A Mixed-Mode and Multi-Level Simulation System", International Symposium on Circuits and Systems, 1991, pp. 2387-2390.

[BRA72] R. K. Brayton, F. G. Gustavson, G. D. Hachtel, "A New Efficient Algorithm for Solving Differential-Algebraic Systems Using Implicit Backward-Differentiation Formulas", Proceedings of the IEEE, Vol. 60, No. 1, Jan. 1972, pp. 98-108.

[BRE72] M. A. Breuer, "A Note on Three-Valued Logic Simulation," IEEE Trans. on Computers, April 1972, pp. 399-402.

[BRE75] M. A. Breuer, Ed., **Digital System Design Automation: Languages, Simulation and Data Base**, Computer Science Press, 1975.

[BRE76] M. A. Breuer and A. D. Friedman, **Diagnosis and Reliable Design of Digital Systems**, Computer Science Press, 1976.

[BRY80] R. E. Bryant, "An Algorithm for MOS Logic Simulation", LAMBDA, 4th Quarter 1980, pp. 46-53.

[BRY84] R. E. Bryant, "A Switch-Level Model and Simulator for MOS Digital Systems", IEEE Trans. on Computers, Vol. c-33, no. 2 , Feb. 1984, pp. 160-177.

[BRY87] R. E. Bryant, "A Survey of Switch-Level Algorithms", IEEE Design and Test of Computers, vol. 4, no. 4, Aug. 1987, pp. 26-40.

[BUR83] J. L. Burns, A. R. Newton, D. O. Pederson, "Active Device Table Lookup Models for Circuit Simulation", International Symposium on Circuits and Systems, Newport Beach, CA, May 1983, pp. 250-253.

[CAR84] C. H. Carlin, A. Vachoux, "On Partitioning for Waveform Relaxation Time-Domain Analysis of VLSI Circuits", International Symposium on Circuits and Systems, Montreal, Canada, May 1984, pp. 701-705.

[CAS88] G. Casinovi, "Macromodeling for the Simulation of Large Scale Analog Integrated Circuit", Ph.D. Dissertation, University of California, Berkeley, August 1988.

[CHA75] B. R. Chawla, H. K. Gummel, and P. Kozak, "MOTIS - An MOS Timing Simulator," IEEE Trans. Circ. and Sys., Vol. 22, 1975, pp. 901-909.

[CHA87] H. P. Chang, J. A. Abraham, "The Complexity of Accurate Logic Simulation", Int. Conf. on Computer-Aided Design, Santa Clara, CA., Nov. 1987, pp. 404-407.

[CHA89] T. Chanak, R. Chadha, , K. Singhal, "Switched-Capacitor Simulation for Full-Chip Verification," Proc. of the Custom Integrated Circuits Conference, San Diego, CA., May 1989, pp. 21.1.1-21.1.4.

[CHA92] Rakesh Chadha, Chandramouli Visweswariah and Chin-Fu Chen, "M^3-A Multilevel Mixed-Mode Mixed A/D Simulator", IEEE Transactions on Computer-Aided Design, Vol. 11, No. 5, May 1992, pp. 575-585.

[CHE84A] C. T. Chen, **Linear System Theory and Design**, Holt, Rinehart and Winston Inc, 1984.

[CHE84B] C. F. Chen, C-Y Lo, H. N. Nham, P. Subramaniam, "The Second Generation MOTIS Mixed-Mode Simulator", Proc. of 21st Design Automation Conference, June 1984, pp. 10-17.

[CHI94] E. Chiprout and M. Nakhla, **Asymptotic Waveform Evaluation and Moment Matching for Interconnect Analysis**, Kluwer Academic Publishers, 1994.

[CHO89] P. Choi and J. Alvin Connelly, "Macromodeling Methodology for Continuous and Discrete Time Transfer Functions", Proceedings of the 32rd Midwest Symposium on Circuits and Systems, August 1989, pp. 997-1000.

[CHU75] L. Chua, P. Lin, **Computer-Aided Analysis of Electronic Circuits: Algorithms and Computational Techniques**, Prentice-Hall, 1975.

[CON93] User's Guide and Reference Manual for CONTECT CAE, CONTEC, April 1993.

[DEG84] A. DeGeus, "SPECS: Simulation Program for Electronic Circuits and Systems," Proc. IEEE Int. Symp. on Circ. and Sys., May 1984, pp. 534-537.

[DEM80] G. De Micheli, " New Algorithms for the Timing Analysis of MOS Circuits" Master Report, University of California, Berkeley, 1980.

[DEM81] G. De Micheli, A. Sangiovanni-Vincentelli, "Numerical Properties of Algorithms for the Timing Analysis of MOS VLSI Circuits", University of California, Berkeley, ERL Memo. UCB/ERL M81/25, May 1981.

[DEM83] G. De Micheli, A. R. Newton, A. Sangiovanni-Vincentelli, "Symmetric Displacement Algorithms for the Timing Analysis for VLSI MOS Circuits", IEEE Trans. on Computer-Aided Design, Vol CAD-2, No. 3, July 1983, pp. 167-180.

[DES69] C. A. Desoer, E. S. Kuh, *Basic Circuit Theory*, McGraw-Hill, 1969.

[DES89] M. P. Desai and I. N. Hajj, "On the Convergence of Block Relaxation Methods for Circuit Simulation", IEEE Transactions Circuits and Systems, Vol. 36, No. 7, July 1989, pp. 948-958.

[DUM86] D. Dumlugol, "Segmented Waveform Relaxation Algorithms for Mixed-Mode Simulation of Digital MOS VLSI Circuits", Ph.D. Dissertation, Katholieke Universiteit Leuven, Oct. 1986.

[EIC65] E. B. Eichelberger, "Hazard Detection in Combinational and Sequential Switching Circuits", IBM J. Res. and Develop., Vol. 9, March 1965, pp. 90-99.

[FAN77] S. P. Fan, M. Y. Hsueh, A. R. Newton and D. O. Pederson, "MOTIS-C A new circuit simulator for MOS LSI circuits," International Symposium on Circuits and Systems, April 1977, pp. 700-707.

[FAN83] S. C. Fang, Y. P. Tsividis, O. Wing, "SWITCAP: A Switched Capacitor Network Analysis Program," IEEE Circuits Syst. Mag., Vol. 5, No. 3, Sept. 1983, pp. 4-10.

[FER91] F. V. Fernandez, A. Rodriguez-Vazquez and J. L. Huertas, "An Advanced Symbolic Analyzer for the Automatic Generation of Analog Circuit Design Equation", IEEE International Symposium on Circuits and Systems, 1991, pp. 810-813.

[FER92] F. V. Fernandez, A. Rodriguez-Vazquez, J. D. Martin and J. L. Huertas, "Accurate Simplification of Large Symbolic Formulae", International Conference on Computer-Aided Design, 1992, pp. 318-321.

[GEA71] C. W. Gear, **Numerical Initial Value Problems in Ordinary Differential Equations**, Prentice-Hall, Englewood Cliffs, N.J., 1971.

[GEA80] C. W. Gear, "Automatic Multirate Methods for Ordinary Differential Equation", Information Processing 80, International Federation of Information Processing, 1980.

[GIE89] G. Gielen, Herman C. C. Walscharts and W. Sansen, "ISAAC: A Symbolic Simulator for Analog Integrated Circuits", IEEE Journal of Solid-State Circuits, Vol. 24, No. 6, Dec. 1989, pp. 1587-1597.

[GIE91] G. Gielen and W. Sansen, **Symbolic Analysis for Automated Design of Analog Integrated Circuits**, Kluwer Academic Publishers, 1991.

[GIE92] G. Gielen, E. Liu, A. Sangiovanni-Vincentelli and P. Gray, "Analog Behavioral Models for Simulation and Synthesis of Mixed-Signal Systems", European Design Automation Conference, 1992, pp. 464-468.

[GOL83] G. H. Golub and C. F. Van Loan, **Matrix Computation**,

Baltimore, MD: Johns Hopkins University Press, 1983.

[GRO87] J. J. Grodstein and T. M. Carter, "SISYPHUS - An Environment for Simulation", Proc. Int. Conf. on CAD, Santa Clara, CA., Nov. 1987, pp. 400-403.

[GYU85] R. S. Gyurscik, "A MOS Transistor Model-Evaluation Attached Processor for Circuit Simulation", Proc. IEEE Int. Conf. on Computer-Aided Design, Santa Clara, CA., Nov. 1985, pp. 234-236.

[HAC71] G. D. Hachtel, R. K. Brayton and F. G. Gustavson, "The Sparse Tableau Approach to Network Analysis and Design", IEEE Trans. on Circ. Theory, Vol. CT-18, Jan. 1971, pp. 101-113.

[HAL88] S. B. Haley, "The Generalized Eigenproblem: Pole-Zero Computation", Proceedings of the IEEE, Vol. 76, No. 2, February 1988, pp. 103-120.

[HAS89] M. M. Hassoun and P. M. Lin, "A New Network Approach to Symbolic Simulation of Large-Scale Networks", IEEE International Symposium on Circuits and Systems, 1989, pp. 806-809.

[HAS91] M. M. Hassoun, "Hierarchical Symbolic Analysis of Large-Scale Systems Using a Mason's Signal Flow Graph Model", IEEE International Symposium on Circuits and Systems, 1991, pp. 802-805.

[HEN85] B. Hennion and P. Senn, "ELDO: A New Third Generation Circuit Simulator Using the One-step Relaxation Method" International Symposium on Circuits and Systems, Kyoto, Japan, June 1985, pp. 1065-1068.

[HIL80] D. Hill, "Multilevel Simulator for Computer-Aided Design", Ph.D. dissertation, Dept. of Elec. Eng., Stanford University, 1980.

[HOS84] B. J. Hosticka, W. Brockherde, U. Kleine and R. Schweer, "Design of Nonlinear Analog Switched-Capacitor Circuits Using Blocks", IEEE Trans. Circuits and Systems, CAS-31, No.4, 1984, pp. 354-368.

[HO75] C. W. Ho, A. E. Ruehli, P. A. Brennan, "The Modified Nodal Approach to Network Analysis", IEEE Trans. on Circ. and Sys., Vol. CAS-22, June 1975, pp. 504-509.

[HSI85] H. Y. Hsieh, A. E. Ruehli, P. Ledak, "Progress on Toggle: A Waveform Relaxation VLSI-MOSFET CAD Program" International Symposium on Circuits and Systems, Kyoto, Japan, June 1985, pp. 213-216.

[HSP92] HSPICE User's Manual, Meta-Software, Inc. 1992.

[HSU93] J.-J. Hsu and C. Sechen, "Low-Frequency Symbolic Analysis of Large Analog Integrated Circuits", IEEE Custom Integrated Circuits Conference, 1993, pp. 14.7.1-14.7.5.

[HUA83] T. Huang, "Analysis of a Method for the Timing Simulation of Large-Scale MOS Circuits Containing Floating Capacitors" Master Report, University of California, Berkeley 1983.

[IEE88] IEEE Computer Society, **IEEE Standard VHDL Language Reference Manual**, The Institute of Electrical and Electronic Engineering, New York, PUBL. NO: IEEE Standards Coordinating Committee 20, 1988.

[INF84] B. Infante, A. Sanders, E. Lock, "Hierarchical Modeling in a Multi-level Simulator", International Conference on Computer-Aided Design, Santa Clara, CA. 1984, pp. 39-41.

[INS84] A. Insinga, "Behavioral Modeling in a Structural Logic Simulator", International Conference on Computer-Aided Design, Santa Clara, CA. 1984, pp. 42-44.

[JOU94] S.-J. Jou, M.-F. Perng, C. C. Su and C. K. Wang, "Hierarchical Techniques for Symbolic Analysis of Large Electronic Circuits", International Symposium on Circuits and Systems, 1994.

[JU91] Y.-C. Ju, V. B. Rao and R. A. Saleh, "Consistency Checking and Optimization of Macromodels", IEEE Transactions on Computer-Aided Design, Vol. 10, No. 8, August 1991, pp. 957-967.

[KAO93] R. Kao and M. Horowitz, "Piecewise Linear Models for Rsim", International Conference on Computer-Aided Design, 1993, pp. 753-758.

[KIM84] Y. Kim, J. E. Kleckner, R. A. Saleh, A. R. Newton, "Electrical-Logic Simulator", International Conference on Computer-Aided Design, Santa Clara, CA., Nov. 1984, pp. 7-10.

[KLE83] J. E. Kleckner, R. A. Saleh, A. R. Newton, "Electrical Consistency in Schematic Simulation", International Conference on Circuits and Computers, NY, October 1983.

[KLE84] J. E. Kleckner, "Advanced Mixed-Mode Simulation Techniques", Ph.D. dissertation, University of California, Berkeley, May 1984.

[KUN86] K. S. Kundert, "Sparse Matrix Techniques and their Application to Circuit Simulation", **Circuit Analysis, Simulation and Design**, A.E. Ruehli, ed., North-Holland Pub. Co., 1986.

[KUN89] K. S. Kundert, J. White and A. Sangiovanni-Vincentelli, "A Mixed Frequency-Time Approach for Distortion Analysis of Switching Filter Circuits", IEEE Journal of Solid-State Circuits, Vol. 24, No. 2, April 1989, pp. 443-451.

[KUN90] K. S. Kundert, J. K. White and A. Sangiovanni-Vincentelli, **Steady-State Methods for Simulating Analog and Microwave Circuits**, Kluwer Academic Publishers, 1990.

[LEE88] E. S. Lee, T-F Fang, "A Mixed-Mode Analog-Digital Simulation Methodology for Full Custom Design", Custom Integrated Circuits Conference, May 1988, pp. 3.5.1-3.5.4.

[LEL82] E. Lelarasmee, A. E. Ruehli, A. L. Sangiovanni-Vincentelli, "The Waveform Relaxation Method for Time-Domain Analysis of Large Scale Integrated Circuits," IEEE Trans. on CAD of IC and Sys., Vol. 1, No. 3, July 1982, pp. 131-145.

[MA92] V. M. Ma, J. Singh and R. Saleh, "Modeling, Simulation and Optimization of Analog Macromodels", IEEE Custom Integrated Circuits Conference, 1992, pp. 12.1.1-12.1.4.

[MAN80] H. De Man, J. Rabaey, G. Arnout, J. Vandewalle, "Practical Implementation of a general computer-aided design technique for switched-capacitor circuits", IEEE Journal of Solid-State Circuits, Vol. sc-15, Apr. 1980, pp. 190-200,

[MAR85] G. Marong and A. Sangiovanni-Vincentelli, "Waveform Relaxation and Dynamic Partitioning for Transient Simulation of Large Scale Bipolar Circuits", International Conference on Computer-Aided Design, Santa Clara, CA, Nov. 1985, pp. 32-35.

[MAS92] S. Masui, "Simulation of Substrate Coupling in Mixed-Signal MOS Circuits", IEEE Symposium on VLSI Circuits Digest of Technical Papers, 1992, pp. 42-42.

[MCC88] W. J. McCalla, **Fundamentals of Circuit Simulation** , Kluwer Academic Publishers, Boston, MA. 1987.

[MHD91] U.S. Army-LABCOM, SLCET-MP, Fort Monmouth, NJ 07703

MHDL Requirements Document, First Edition, 1991.

[MIC94] The Design Center Circuit Analysis User's Guide, Microsim Corp., Jan. 1994.

[MOL78] C. B. Moler, C. F. Van Loan, "Nineteen Dubious Ways to Compute the Exponential of a Matrix", SIAM Review 20, 1978, pp. 801-836.

[NAG75] L.W. Nagel, "SPICE2: A Computer Program to Simulate Semiconductor Circuits," Electronics Research Laboratory Rep. No. ERL-M520, University of California, Berkeley, May 1975.

[NAG80] L. W. Nagel, "ADVICE for Circuit Simulation," International Symposium on Circuits and Systems, May 1980.

[NEW77] A. R. Newton, D. O. Pederson, "Analysis Time, Accuracy and Memory Tradeoffs in SPICE2", 12th Asilomar Conference on Circuits, Systems and Computers, Asilomar CA, November 1977, pp. 6-9.

[NEW78] A. R. Newton, "The Simulation of Large-Scale Integrated Circuits", Ph.D. dissertation, University of California, Berkeley, ERL Memo. ERL-M78/52, July 1978.

[NEW79] A. R. Newton, "The Analysis of Floating Capacitors for Timing Simulation," Proc. 13th Asilomar Conf. on Circ., Sys. and Comp., Asilomar CA, November 1979, pp. 433-436.

[NEW81] A. R. Newton, "Timing, Logic and Mixed-Mode Simulation for Large MOS Integrated Circuits", in **Computer-Aids for VLSI Circuits**, Sijthoff & Noordhoff International Publishers, The Hague, 1981, pp. 175-239.

[NEW83] A. R. Newton, A. Sangiovanni-Vincentelli, "Relaxation-based Circuit Simulation", IEEE Trans. on Elec. Dev., Vol. ED-30, No. 9, Sept. 1983, pp. 1184-1207.

[ODR86] P. Odryna, K. Nazareth, C. Christensen, "A Workstation-based Mixed-Mode Circuit Simulator", Proc. of the 23rd Design Automation Conference, June 1986, pp. 186-191.

[ORT70] J. M. Ortega and W. C. Rheinbolt, **Iterative Solution of Nonlinear Equations in Several Variables**, Academic Press, 1970.

[OVE88] D. Overhauser, I. Hajj, "A Tabular Macromodelling Approach to Fast Timing Simulation Including Parasitics," International Conference on

Computer-Aided Design, Santa Clara, CA., 1988, pp. 70-73.

[OVE89A] D. Overhauser, I. Hajj, "Automatic Mixed-Mode Timing Simulation", International Conference on Computer-Aided Design, 1989, pp. 84-87.

[OVE89B] D. Overhauser, "Fast Timing Simulation of MOS VLSI Circuits", Ph.D. Dissertation, University of Illinois, Aug. 1989.

[PEN81] P. Penfield, J. Rubenstein, "Signal Delays in RC Tree Networks," Proc. of 18th Design Automation Conference, June 1981, pp. 613-617.

[PIL90] L. T. Pillage and R. A. Rohrer, "Asymptotic Waveform Evaluation for Timing Analysis", IEEE Transactions on Computer-Aided Design, Vol. 9, No. 4, April 1990, pp. 352-366.

[PSP90] PSPICE Users Manual, Microsim Corp., Jan. 1990.

[QUA89] T. Quarles, "Analysis of Performance and Convergence Issues for Circuit Simulation," Ph.D. Dissertation, UCB/ERL M89/42, University of California, Berkeley, April 1989. Berkeley, CA. 1989.

[RAO85] V. Rao, "Switch-level Timing Simulation of MOS VLSI Circuits", Ph.D. dissertation, University of Illinois, UILU-ENG-85-2207, R-1032, Jan. 1985.

[RAO89] V. Rao, D. Overhauser, I. Hajj, T. Trick, **Switch-level Timing Simulation of MOS VLSI Circuits**, Kluwer Academic Publishers, Boston, MA., 1989.

[RAB79] N. B. G. Rabbat, A. Sangiovanni-Vincentelli and H. Y. Hsieh, "A Multilevel Newton Algorithm with Macromodelling and Latency for the Analysis of Large-Scale Nonlinear Circuits in the Time Domain", IEEE Trans. on Circ. and Sys., Vol. CAS-26, Sept. 1979, pp. 733-741.

[RAG93] V. Raghavan, R. A. Rohrer, L. T. Pillage, J. Y. Lee, J. E. Bracken and M. M. Alaybeyi, "AWE-Inspired", IEEE Custom Integrated Circuits Conference, 1993, pp. 18.1.1-18.1.8.

[RUM89] M. Rumsey and J. Sackett, "An ASIC Methodology for Mixed Analog-Digital Simulation", 26th ACM/IEEE Design Automation Conference, 1989, pp. 618-621.

[RUT93] R. Rutenbar, "Analog Design Automation: Where are We?

Where are We Going?", IEEE Custom Integrated Circuits Conference, 1993, pp. 13.1.1-13.1.8.

[SAK80] K. Sakallah and S. W. Director, "An Activity-Directed Circuit Simulation Algorithm," International Conference on Circuits and Computers, October 1980.

[SAK81] K. A. Sakallah, "Mixed Simulation of Electronic Integrated Circuits", Ph.D. dissertation, Carnegie-Mellon University, DRC-02-07-81, Nov. 1981.

[SAK85A] K. Sakallah, "Polynomial Terminal Equivalent Circuits as Dormant Models in Event Driven Circuit Simulation", International Conference on Computer-Aided Design, Santa Clara, CA, 1985, pp. 179-181.

[SAK85B] K. Sakui, T. Shima, T. Hayashi, Fumio Horiguchi and Mitsugi Ogura, "A Simplified Accurate Three-Dimensional Table Look-Up MOSFET Model for VLSI Circuit Simulation", Custom Integrated Circuits Conference, 1985, pp. 347-351.

[SAL83] R. A. Saleh, J. E. Kleckner and A. R. Newton, "Iterated Timing Analysis and SPLICE1", International Conference on Computer-Aided Design, Santa Clara, CA., 1983, pp. 139-140.

[SAL84] R. Saleh, "Iterated Timing Analysis and SPLICE1", Master Report, University of California, Berkeley, 1984.

[SAL89A] R. Saleh, "iSPLICE3 User's Guide", University of Illinois, 1992.

[SAL89B] R. Saleh, A. R. Newton, "The Exploitation of Latency and Multirate Behavior using Nonlinear Relaxation for Circuit Simulation," IEEE Trans. on Computer-Aided Design of Circ. and Sys., Dec. 1989, pp. 1286-1298.

[SAL94A] R. A. Saleh, D. L. Rhodes, E. Christen and B. A. A. Antao, "Analog Hardware Description Languages", Custom Integrated Circuits Conference, May 1994.

[SAL94B] R. A. Saleh, S. J. Jou, D. Overhauser, X. Xu and Y. Wang, " Benchmark Circuits for Mixed-Mode Simulation" Custom Integrated Circuits Conference, 1994.

[SAN77] A. Sangiovanni-Vincentelli, L. K. Chen and L. O. Chua, "A New Tearing Approach-Node Tearing Nodal Analysis", International

Symposium on Circuits and Systems, 1977, pp. 143-147.

[SAV93] P. Saviz and O. Wing, "Circuit Simulation by Hierarchical Waveform Relaxation", IEEE Transactions on Computer-Aided Design of Integrated Circuits and Systems, Vol. 12, No. 6, June 1993, pp. 845-860.

[SED88] S. Seda, M. Degrauwe and W. Fichtner, "A Symbolic Analysis Tool for Analog Circuit Design Automation", IEEE International Conference on Computer-Aided Design, Nov. 1988, pp. 488-491.

[SED92] S. Seda, M. Degrauwe and W. Fichtner, "Lazy-Expansion Symbolic Expression Approximation in SYNAP", International Conference on Computer-Aided Design, 1992, pp. 310-317.

[SHE88] J. Sherred, "A Phase-Locked Clock Generator for VLSI Applications", M. S. Thesis, Mass. Inst. of Tech, June 1988.

[SHI82] T. Shima, T. Sugawara, S. Moriyama and H. Yamada, "Three-Dimensional Table Look-up MOSFET Model for Precise Circuit Simulation", IEEE Journal of Solid-State Circuits, Vol. Sc-17, No. 3, June 1982, pp. 449-454.

[SHI83] T. Shima, H. Yamada and R. L. M. Dang, "Table Look-up MOSFET Modeling System Using 2-D Device Simulator and Montonic Piecewise Cubic Interpolation", IEEE Trans. CAD, Vol.2, No.2, April 1983, pp. 121-126.

[SIN91] J. Singh and R. Saleh, "iMACSIM: A Program for Multi-Level Analog Circuit Simulation", International Conference on Computer-Aided Design, 1991, pp. 16-19.

[SIN94A] J. Singh, X. Xu and R. Saleh, "Simulator Characteristics Needed to Interface to Analog HDLs", International Conference on Simulation and Hardware Description Languages (ICSHDL), Tempe, AZ., Jan. 1994, pp. 31-36.

[SIN94B] J. Singh, "Techiques for Analog Multilevel Simulation", Ph.D Thesis, University of Illinois at Urbana-Champaign, 1994.

[SPA88] R. Sparkes, G. Boyle and R. Woolhiser, "Evaluation of Macro Models for Mixed Analog/Digital Circuits", IEEE Custom Integrated Circuits Conference, 1988, pp. 3.4.1-3.4.6.

[SUN81] Y. Sun, "Direct Analysis of Time-Varying Continuous and Discrete Difference Equations with Application to Nonuniformly

Switched-Capacitor Circuits", IEEE Transation Circuits and Systems, Vol. CAS-28, No. 2, Feb. 1981, pp. 93-100.

[SUY89] K. Suyama, "Analysis, Simulation, and Application of Linear and Nonlinear Switched-Capacitor and Mixed Switched-Capacitor/Digital Networks", Ph.D. Thesis, Columbia University, 1989.

[SZY75] S. A. Szygenda and E. W. Thompson, "Digital Logic Simulation in a Time-Based, Table-Driven Environment. Part 1. Design Verification," IEEE Computer Magazine, March 1975, pp. 24-36.

[SWI89] SWITCAP-II Users Guide, Columbia University, 1989.

[TAH89] H. El Tahawy, A. Chianale and B. Hennion, "Functional Verification of Analog Blocks in Fideldo: A Unified Mixed-Mode Simulation Environment", International Symposium on Circuits and Systems, 1989, pp. 2012-2015.

[TEM78] G. C. Temes, H. J. Orchard and M. Jahanbegloo, "Switched-Capacitor Filter Design Using the Bilinear Z-Transform", IEEE Transation Circuits and Systems, Vol. CAS-25, No. 12, Dec. 1978, pp. 1038-1044.

[TER83] C. Terman, "RSIM - A Logic-Level Timing Simulator", Int. Conf. on Comp. Design, Port Chester, NY, 1983, pp. 437-440.

[THA92] T. J. Thatcher and Resve A. Saleh, "Automatic Partitioning and Dynamic Mixed -Mode Simulation", IEEE Custom Integrated Circuits Conference, 1992, pp. 12.7.1-12.7.4.

[THO91] D. E. Thomas and Philip R. Moorby, **The Verilog Hardware Description Language**, Kluwer Academic Publishers, Boston, 1991.

[TRI90] R. Trihy and C. Lyden, "An Accurate and Compact Model for the Transient Simulation of Continuous-Time", IEEE Custom Integrated Circuits Conference, May 1990, pp. 8.8.1-8.8.4.

[TSA85] D. Tsao, C-F Chen, "A Fast Timing Simulation for Digital MOS Circuits", International Conference on Computer-Aided Design, Santa Clara, CA. Nov. 1985, pp. 185-187.

[UYE88] J. P. Uyemura, **Fundamentals of MOS Digital Integrated Circuits**, New York: Addison-Wesley Publishing Company, 1988.

[VAR62] R. S. Varga, **Matrix Iterative Analysis**, Prentice-Hall, 1962.

[VER93] N. Verghese and D. Allstot, "Rapid Simulation of Substrate

Coupling Effects in Mixed-Mode ICs", IEEE Custom Integrated Circuits Conference, 1993, pp. 18.3.1-18.3.4.

[VID86] L. Vidigal, S. Nassif, S. Director, "CINNAMON: Coupled Integration and Nodal Analysis of MOS Networks," 23rd Design Automation Conference, June 1986, pp. 179-185.

[VIS86] C. Visweswariah, "SPECS2: A Timing Simulator", M.S. Report, Carnegie-Mellon University, Report No. CMUCAD-86-24, October 1986.

[VIS88] C. Visweswariah, R. Chadha and C.-F. Chen, "Model Development and Verification for High Level Analog Blocks", 25th ACM/IEEE Design Automation Conference 1988, pp. 376-382.

[VHD91] VHDL Analog Sup-PAR Group, Minutes of the October Meeting San Francisco, CA. 1991.

[VLA83] J. Vlach and K. Singhal, **Computer Methods for Circuit Analysis and Design**, Van Nostrand Reinhold Electrical/Computer Science and Engineering Series, 1983.

[VLA90] M. Vlach, "Modeling and Simulation with Saber", IEEE ASIC Design Conference, 1990, pp. T-11.1-T-11.9.

[WAL86] U. V. Wali, R. N. Pal and B. Chatterjee, "Compact Modified Nodal Approach for Switched-Capacitor Network Analysis", IEEE Transactions on Computer-Aided Design, Vol. 5, No. 3, July 1986, pp. 443-447.

[WAR78] D. E. Ward and R. W. Dutton, "A Charge-Oriented Model for MOS Transient Capacitances", IEEE J. Solid-state Circuits, Vol. SC-13, Oct. 1978, pp. 703-707.

[WEE73] W. T. Weeks, A. J. Jimenez, G. W. Mahoney, D. Mehta, H. Qassemzadeh, and T. R. Scott, "Algorithms for ASTAP -- A Network Analysis Program," IEEE Trans. on Circuit Theory, Vol. CT-20, No. 6, November 1973, pp. 628-634.

[WES85] N. Weste and K. Eshraghian, **Principles of CMOS VLSI Design**, New York: Addision Wesley Publishing Company, 1985.

[WHI83] J. White and A. Sangiovanni-Vincentelli, "RELAX2: A New Waveform Relaxation Approach for the Analysis of LSI MOS Circuits", International Symposium on Circuits and Systems, Newport Beach, May 1983, pp. 756-759.

[WHI84] J. White and A. Sangiovanni-Vincentelli, "RELAX2.1 - A Waveform Relaxation Based Circuit Simulation Program" Custom Integrated Circuits Conference, Rochester, New York, June 1984, pp. 232-236.

[WHI85A] J. White, A. L. Sangiovanni-Vincentelli, "Partitioning Algorithms and Parallel Implementation of Waveform Relaxation Algorithms for Circuit Simulation", International Symposium on Circuits and Systems, Kyoto, Japan, June 1985, pp. 221-224.

[WHI85B] J. White, R. Saleh, A. Sangiovanni-Vincentelli, A. R. Newton "Accelerating Relaxation Algorithms using Waveform-Newton, Step Refinement and Parallel Techniques," International Conference on Computer-Aided Design, Santa Clara, CA, Nov. 1985, pp. 5-7.

[WHI85C] J. White, "The Multirate Integration Properties of Waveform Relaxation, with Application to Circuit Simulation and Parallel Computation", Ph.D. dissertation, University of California, Berkeley, ERL Memo. No. UCB/ERL 85/90, Nov. 1985.

[WHI86] J. White and A. Sangiovanni-Vincentelli, **Relaxation Techniques for the Simulation of MOS VLSI Circuits**, Kluwer Academic Publishers, Boston, MA., 1987.

[WIL92] L. A. Williams III and B. A. Wooley, "MIDAS-A Functional Simulator for Mixed Digital and Analog Sampled Data Systems", International Symposium on Circuits and Systems, 1992, pp. 2148-2151.

[WOL91] S. Wolfram. Mathematica A System for Doing Mathematics by Computer. Addison-Wesley Inc., New York, 1991.

[YAN80] P. Yang, I. N. Hajj and T. N. Trick, "SLATE: A Circuit Simulation Program with Latency Exploitation and Node Tearing", International Conference on Circuits and Computers, October 1980.

[YAN83] P. Yang, B. D. Epler, P. K. Chatterjee, "An Investigation of the Charge Conservation Problem for MOSFET Circuit Simulation", IEEE Journal of Solid-State Circuits, Vol. SC-18, No.1, Feb. 1983, pp. 128-138.

INDEX

ABOUT THE AUTHORS

Resve A. Saleh obtained his B. Eng. Degree (Electrical) from Carleton University, Ottawa, Canada, in 1979, and his M.S. and Ph.D. degrees from U.C. Berkeley in 1983 and 1986, respectively. He has worked in industry for Mitel Corporation, Tektronix Inc., and Toshiba Corporation. He joined the University of Illinois in 1986 where he is currently an Associate Professor in Electrical and Computer Engineering directing research in mixed-mode simulation, analog multilevel simulation, and parallel processing. He served as the technical program chair of the Custom Integrated Circuits Conference in 1993, and has served on the technical committees of the Design Automation Conference, the MidWest Symposium on Circuits and Systems, the International Conference on Computer Design. He is currently an Associate Editor with the IEEE Transactions on Computer-Aided Design, and was was also an editor for the Simulation column of Circuits and Devices Magazine. He is serving as the chairman of the IEEE SCC-30 committee involved in standardizing an Analog Hardware Description Language.

Shyh-Jye Jou obtained his B.S. degree in Electrical Engineering from National Chen-Kung University in 1982, and M. S. and Ph.D degrees in Electronics from National Chiao-Tung University, Taiwan, Republic of China in 1984 and 1988, respectively. Currently, he is an Associate Professor of the Electrical Engineering at National Central University, Taiwan, Republic of China. He was a visiting research Associate Professor in the Coordinated Science Laboratory at University of Illinois during the 1993-94 academic year. He served on the technical committee of the 1994 Custom Integrated Circuits Conference. His research interests include computer aided-design tools of integrated circuits; specifically, simulation and VLSI digital circuit design.

A. Richard Newton received the B. Eng. (elect.) and M. Eng. Sci. degrees from the University of Melbourne, Melbourne, Australia, in 1973 and 1975, respectively, and the Ph.D. degree from the University of California, Berkeley, in 1978. He is currently a Professor and Vice Chairman of the Department of Electrical Engineering and Computer Sciences, University of California, Berkeley. He was the Technical Program Chairman of the 1988 ACM/IEEE Design Automation Conferences, and a consultant to a number of companies for computer-aided design of integrated circuits. His research interests include all aspects of the computer-aided design of integrated circuits, with emphasis on simulation, automated layout techniques, and design methods for VLSI integrated circuits. Dr. Newton was selected in 1987 as the national recipient of the C. Holmes McDonald Outstanding Young Professor Award of Eta Kappa Nu.

MIX
Papier aus verantwortungsvollen Quellen
Paper from responsible sources
FSC® C105338

If you have any concerns about our products,
you can contact us on
ProductSafety@springernature.com

In case Publisher is established outside the EU,
the EU authorized representative is:
Springer Nature Customer Service Center GmbH
Europaplatz 3, 69115 Heidelberg, Germany

Printed by Libri Plureos GmbH
in Hamburg, Germany